Wartime and Aftermath

English Literature and its Background
1939–60

BERNARD BERGONZI

Oxford · New York
OXFORD UNIVERSITY PRESS
1993

Oxford University Press, Walton Street, Oxford OX2 6DP

Oxford New York Toronto
Delhi Bombay Calcutta Madras Karachi
Kuala Lumpur Singapore Hong Kong Tokyo
Nairobi Dar es Salaam Cape Town
Melbourne Auckland Madrid

and associated companies in
Berlin Ibadan

Oxford is a trade mark of Oxford University Press

British Library Cataloguing in Publication Data
Data available

Library of Congress Cataloging in Publication Data
Bergonzi, Bernard.
Wartime and aftermath: English literature and its background,
1939–60 / Bernard Bergonzi.
p.   cm.   "An OPUS book"—CIP galley.
Includes bibliographical references.
1. English literature—20th century—History and criticism.
2. World War, 1939–1945—Great Britain—Literature and the war.
3. World War, 1939–1945—Great Britain—Influence.   4. War stories,
English—History and criticism.   5. War poetry, English—History and
criticism.   I. Title.
820.9'358—dc20   PR478.W67B46   1993   92–26593
ISBN 0–19–219242–6
ISBN 0–19–289222–3 Pbk

1 3 5 7 9 10 8 6 4 2

Typeset by Best-set Typesetter Ltd, Hong Kong
Printed in Great Britain by
Biddles Ltd.
Guildford and King's Lynn

Preface

This book is part of a series with the general title 'English Literature and its Background'. I interpret the latter word as 'context', which is likely to vary from one text to another, rather than as a fixed, unchanging entity standing behind the literary action. In the present study, 'background' is frequently historical and political, and in the earlier part of the period involves the demands made on writers by the most devastating war in history; but I also see it as a matter of ideas, and sometimes of other literature; one text can well act as the 'background' to another. My approach is, then, eclectic, with no conviction that socio-economic factors are necessarily more fundamental than intellectual, cultural or religious ones.

The book opens in 1939, a year proposed by history rather than literature; the first shots of the Second World War were fired on 1 September, and I have chosen to begin the story at that point, avoiding any preliminary discussion of the run-up to war. It would have been a problem to know where to start, and professional historians are in any case divided on such matters as the real condition of England in the 1930s, or the causes of the Second World War. The year 1939 is of some literary significance, as the year in which Yeats and Freud and Ford Madox Ford died, and W. H. Auden left England for America, in a symbolically resonant departure. For many English writers it was the year in which, as they later recalled, the shaky order of *l'entre deux guerres* finally came to pieces.

Endings are harder to handle than beginnings, and decades are fictional constructions; my choice of 1960 as a closing date may seem arbitrary, but I believe there are good reasons for ending at that point, as I go on to indicate. My treatment of the period shows six years of war being followed by fifteen years of slow recovery. Closure in 1960 means that discussion of the writers who emerged in the 1950s is cut short early in their careers, but I hope there is some compensating interest in seeing their interactions in a cross-section of the period.

The other noun in the subtitle of the book, 'literature', has become problematical in recent years, so I had better say—odd though it seems to have to—that I think that there is such a thing as literature. It is continually and inextricably caught up in historical and cultural factors, but it remains distinct from them; it is, in Ezra Pound's great formulation, 'news which stays news'. One problem in writing literary history is a pull towards cultural history, or even cultural antiquarianism, so that writing about the poetry of a period becomes not very different from writing about, say, its gardening manuals. I find cultural history fascinating, but I believe that literary history has rather different aims, and inevitably involves one in writing literary criticism. I have tried to give enough space to the works I discuss to say something relevant about them, which includes saying how good I think they are. I have tried to avoid making the great the enemy of the good; the 1940s and 1950s were not, perhaps, a period of great literature, compared with the years of modernist achievement earlier in the century, but, as I try to show, there was much more good work written then, in a variety of modes, than is sometimes thought.

I have tried (not with complete success) to avoid the kind of literary history which in the interests of including everyone declines into strings of names with a summary label attached. Since space is restricted, I have had to be selective, and there are omissions of writers who might have been discussed in a longer book. I do not believe I have left out anyone whom I take to be really important, though I accept that 'importance' is another elusive term, and is not necessarily the same as merit. Some writing I am not enthusiastic about is discussed because it is inescapably part of the background, or the climate, of the time. For the most part I have concentrated on writing which aims to be literary, whether or not it succeeds; without doubt, a study of the sub-literary and formulaic writing of this period would be of great socio-cultural interest. But that is a task for someone else.

I was a schoolboy during the war years and a young man in the 1950s, and much of the pleasure I have found in writing this book has been in revisiting texts that I read when they first came out. A few I had quite forgotten and had to rediscover, and some I missed originally and have now read for the first time. I hope that readers of my generation will enjoy sharing the *recherche*, and that younger ones will find new sources of interest and satisfaction.

Contents

I

Blackout to Blitz

The first shots of the Second World War were fired at dawn on 1 September 1939, as the German army invaded Poland. Later that day, W. H. Auden, the most admired and influential English poet of his generation, sat in a New York bar, pondering on fate and history. He preserved the experience in his poem, 'September 1, 1939', presenting himself as uncertain and afraid, 'as the clever hopes expire | Of a low dishonest decade.' In that magisterially dismissive phrase he severed his connection with the 1930s, the period which had produced his best and most memorable poetry. The poem marks a deep historical divide; but it also indicates Auden's personal shift from the Marxism laced with psychoanalysis that had so affected his poetry of the Thirties to a form of existentialist Protestantism. The poem has long been a favourite anthology piece, but Auden was always unhappy with it, particularly with its most famous line, 'We must love one another or die', which he came to regard as dishonest; he tried changing it to 'We must love one another and die', but eventually he excised the whole poem from his canon.

Britain and France were committed by treaty to come to the defence of Poland, but they took their time. An ultimatum was sent to Hitler demanding the withdrawal of German forces by 3 September. It was ignored, and at 11 a.m. on that day the British prime minister, Neville Chamberlain, made a lugubrious broadcast to the nation, announcing a state of war with Germany. Shortly afterwards the air raid sirens sounded in London. It proved to be a false alarm, but it seemed briefly as if the Nazi war machine had moved into action with breathtaking efficiency and that the aerial bombardment of the capital, feared and expected for several years, was about to begin. People long remembered what they were doing on that day; Rob Wilton, a popular radio comedian, used to deliver a monologue which began, 'The day war broke out . . .' Chamberlain's broadcast and the subsequent air raid warning appear in two of the best novels of the war years: at the beginning of Evelyn Waugh's *Put Out More Flags* (1942), and at the end of Patrick Hamilton's *Hangover Square* (1941).

The poet Roy Fuller was alone at his home in Blackheath on 3 September; his wife and 2-year-old son had moved for safety to her parents' home at Blackpool two days before. He had dug a shallow slit trench in his back garden for protection against air raids, though his wife said it looked as if he had dug his own grave; when the sirens sounded on the morning of the 3rd he went and lay self-consciously in it. No raiders came and the 'all clear' was soon sounded. The day seemed deceptively like any other fine Sunday at the end of summer, and Fuller later visited his friend and fellow-poet Julian Symons at Denmark Hill, a few miles away in the sprawling suburbs of South London.[1] Two younger poets, Derek Stanford and John Bayliss, spent that Sunday evening larking with a couple of girls by the Thames at Kew. Taking a dimly lit late trolley-bus home through the already blacked-out streets they were disturbed by intense rumbles of thunder, which they persuaded themselves were gunfire and the prelude to a Nazi invasion.[2] Their imaginations had been influenced by Alexander Korda's famous film of H. G. Wells's *The Shape of Things to Come*, made in 1936, which correctly located the bombing of London in 1940, and which provided pervasive images of the likely shape of a future war.

Two undergraduates, still on vacation, returned to Oxford. Richard Hillary had spent two idle pleasant years, trying to combine athleticism and aestheticism, wanting to be a writer and spending much of his time rowing. He welcomed the outbreak of war, as providing a goal and purpose in life: 'the war solved all problems of a career, and promised a chance of self-realization that would normally take years to achieve. As a fighter pilot I hoped for a concentration of amusement, fear and exaltation which it would be impossible to experience in any other form of existence.'[3] Hillary was a member of the

University Air Squadron, and on 3 September he reported to the Volunteer Reserve Centre in Oxford, only to be told that his services would not yet be required. Keith Douglas, like Hillary, was a public-school boy who combined literary and aesthetic interests with a taste for violent action. He and a friend listened to Chamberlain's broadcast in silence, looking out across a Sussex landscape. Then Douglas travelled to Oxford to enlist. He let it be known that he would join a good cavalry regiment and 'bloody well make my mark in this war. For I will not come back.' Passing a memorial to the dead of the First World War he casually remarked that his name would be on the next one.[4] He spent three more terms at Oxford before being called up.

Hillary and Douglas became well known as writers, and neither survived the war. Hillary went through the Battle of Britain as a fighter pilot, was shot down and appallingly burned. He recovered, and in 1942 published *The Last Enemy*, a personal record of action, and of the state of mind of a representative member of his generation; it quickly became a bestseller. Hillary went back to flying and was accidently killed on a night training flight. Douglas's reputation was slower to develop, but he is now generally regarded as the best English poet of the Second World War. He went through the North African campaign as a tank commander, and described his experiences in a vivid prose memoir, *Alamein to Zem Zem* (1946). Although he published some poems during his lifetime, neither *Alamein to Zem Zem* nor his collected poems appeared until after the war. Douglas was killed in the fighting in Normandy in 1944 at the age of 24.

After the fears and excitement of the first day of war, came a sense of anticlimax. There was no bombing, and material existence continued much as before, apart from the blackout and the disruption of family life caused by the evacuation of millions of children from large cities. Roy Fuller's poem 'Autumn 1939' begins:

> Cigar-coloured bracken, the gloom between the trees,
> The straight wet by-pass through the shaven clover,
> Smell of the war, as if already these
>   Were salient or cover.

In Europe, Britain and France were powerless to help the Poles, being in no military condition to launch an invasion of Germany.

Poland was defeated within a month and the nation was divided, in its fourth partition in history, between Nazi Germany and Soviet Russia, ideological enemies who had astonished the world in August by uniting in a non-aggression pact. September 1939 to April 1940 was the period of the 'phoney war' or, as Evelyn Waugh described it in the dedicatory letter of *Put Out More Flags*, 'that odd, dead period before the Churchillian renaissance, which people called at the time the Great Bore War'. The French army and the small British expeditionary force remained in a defensive posture behind the supposedly impregnable fortifications known as the Maginot Line, while the Germans did the same behind their equivalent, the Siegfried Line (though the British claimed in a popular song that they were going to hang out their washing on the Siegfried Line). To a population which had been expecting Armageddon the Phoney War was a perplexing time, where the worst hazards came from navigating the blacked-out streets. *Put Out More Flags* provides a satirical account of the early days, when writers and intellectuals flocked to join the ever-expanding Ministry of Information, while upper-class young men pulled strings to get commissions in the right regiments. Waugh himself had been part of this process, but he presented his own experience with sardonic detachment. He later gave a more extended account of this phase in *Sword of Honour*, as did Anthony Powell in *The Valley of Bones*.

Easter 1940 saw the publication of a major poem inspired by the war, T. S. Eliot's *East Coker*, which became the second of his *Four Quartets*. In the late Thirties Eliot believed that his future lay in drama—a belief which had been strengthened by the modest success of *The Family Reunion* on the stage in 1939—and that he had finished with 'pure' poetry with *Burnt Norton* in 1936. But the war closed the theatres for a time; and the blackout and anxiety about the future turned public consciousness inward. Eliot shared this prevalent introspection, and early in 1940 he wrote a new poem in the lyrical-meditative manner of *Burnt Norton*, which he used as a formal model. *East Coker* starts in the Somerset village from which Andrew Eliot had set out for America at the end of the seventeenth century, and it ends with the words, 'In my end is my beginning.' It is a personal poem in which Eliot, the great expounder of poetic impersonality, but also an adoptive Englishman, looks back to his remote family origins and reflects on the problems of his art:

> So here I am, in the middle way, having had twenty years—
> Twenty years largely wasted, the years of *l'entre deux guerres*
> Trying to use words, and every attempt
> Is a wholly new start and a different kind of failure . . .

Though *East Coker* is a personal poem, it had a wide resonance. There was a conviction that public men and institutions—'the captains, merchant bankers, eminent men of letters, | The generous patrons of art, the statesmen and rulers'—were indeed in the dark, and had led the country into it. 'They all go into the dark' Eliot wrote at the beginning of section III, after invoking the 'dark, dark, dark' of Milton's Samson. And so did his readers, stumbling in the blackout. The words 'dark' and 'darkness' echo throughout 'East Coker'. The poem was in tune with a general mood. It appeared in the Easter number of *The New English Weekly* and was instantly acclaimed. Then it was reprinted as a pamphlet and, remarkably for a 'difficult' modern poem, sold several thousand copies. This response encouraged Eliot to continue with what eventually became *Four Quartets*.

At the beginning of 1940 Cyril Connolly launched a new monthly magazine, *Horizon*, which was intended to uphold and defend the ideals of high culture in wartime England (it appeared just in time; soon afterwards there was a ban on new periodicals, in order to save paper). Like many writers of the 1930s, Connolly combined aesthetic inclinations with loosely left-wing attitudes. His *Enemies of Promise* had been justly praised on its publication in 1938. In that book, Connolly made a perceptive and witty but dispirited analysis of the

contemporary literary situation, and looked back over his
school-days and early life, tracing his own development as a
specimen of the modern man of letters. Connolly was highly
intelligent and widely read, but he was the worst enemy of
his own promise, trapped in sloth and a self-pitying and self-
defeating state of mind. Nevertheless, *Horizon* played a
major part in the cultural life of the war years. It took an
unashamedly élitist line about art, and was often attacked
for its cultural snobbery and its seeming indifference to the
war effort; yet it had devoted readers and defenders, including many in the armed forces, for whom *Horizon* provided
nostalgic reminders of higher things amid the dangers and
privations of their lives. To this extent *Horizon*, despite the
often gloomy and querulous tone of Connolly's editorials,
could claim to strengthen morale.

   Connolly and Evelyn Waugh had enjoyed a complicated
friendship since they were at Oxford together. Waugh was
fond of Connolly in a personal way, but he had little respect
for him as a writer, and he lost no opportunity for teasing and
needling him, as in the figure of General Connolly in *Black
Mischief* or the grotesque Connolly family of evacuee children
in *Put Out More Flags*. In that novel, *Horizon*, suddenly
manifesting itself during the Phoney War, is satirized as
The Ivory Tower; it appears again as *Survival* in *Sword of
Honour*, where Connolly himself is presented in the guise of
the domineering aesthete, Everard Spruce. Yet their friendship survived the mockery, and in 1948 Connolly devoted a
whole issue of the magazine to the first publication of Waugh's
The Loved One. Though *Horizon* upheld aesthetic ideals, it
was catholic in its range of contributors, and by no means
all of them were aesthetes. One regular contributor was
the dissident, anti-Marxist socialist George Orwell, whom
Connolly had known at Eton. Though Orwell had written a
number of novels before the war, his strengths as a writer were
most evident in works of personal narrative, such as *The Road
to Wigan Pier* and *Homage to Catalonia*; the latter is a classic
of war literature in which Orwell recounts his experiences in
the Spanish Civil War, where he fought in a Trotskyist militia

and was badly wounded. It made Orwell *persona non grata* among many of the English Left by favouring the Trotskyists and Anarchists against the Stalinists in the bitter dissensions that divided the Republican side.

Horizon provided Orwell with a platform in which he could develop his characteristic blend of political pamphleteering and literary and social criticism. His collection of essays, *Inside the Whale* (1940), launched him as an outstanding commentator on politics, literature, and culture. It included an essay on boys' weeklies, first published in *Horizon*, which was a pioneering study in the analysis of popular culture. The severe Cambridge critic Q. D. Leavis, who had also worked in that area (though lacking Orwell's empathy with the material), wrote an approving review in *Scrutiny*, describing Orwell as 'a live mind working through literature, life and ideas . . . he is that rare thing, a non-literary writer who is also sensitive to literature'.

The Phoney War came to an abrupt end on 9 April 1940, when the German Army rapidly occupied Denmark, and invaded Norway. The British and French, who themselves had been no respecters of Norwegian neutrality, gave what military assistance they could, and for the first time British troops were seriously in action. The Norwegian campaign was an ignominious disaster for the Allies, and the Germans soon occupied the whole country. The fighting in Norway appears near the end of *Put Out More Flags*, marking a decisive end to the futilities that Waugh has satirically described, and killing off one of his characters now in uniform, the ineffectual aesthete and cuckold, Cedric Lyne. David Gascoyne's poem, 'A Wartime Dawn' ends its painful evocation of insomnia with a sharp transition from personal sensations to the new collective experience of war:

> Now head sinks into pillows in retreat
> Before this morning's hovering advance;
> (Behind loose lids, in sleep's warm porch, half hears
> White hollow clink of bottles,—dragging crunch
> Of milk-cart wheels,—presently a snatch
> Of windy whistling as the newsboy's bike winds near,

> Distributing to neighbour's peaceful steps
> Reports of last-night's battles); at last sleeps.
> While early guns on Norway's bitter coast
> Where faceless troops are landing, renew fire:
> And one more day of War starts everywhere.

After the occupation of Norway the German army launched a brilliantly successful offensive in the West, a *Blitzkrieg* or lightning war. Belgium and Holland were defeated in a few days, the Maginot Line was outflanked and the Germans thrust deep into France. Paris was occupied on 13 June, and on the 21st the French signed an armistice. Britain faced invasion, but Churchill, who had replaced Chamberlain as prime minister in May, resolved to fight on. Meanwhile, the bulk of the British army in Flanders, minus nearly all their equipment, had been evacuated through the port of Dunkirk and brought back to England. In popular mythology 'Dunkirk' is remembered as a glorious quasi-victory; it was indeed an extraordinarily lucky, even miraculous, escape, helped by fine weather and calm seas and the fact that the Germans inexplicably halted their advance on the port. Nevertheless, Churchill privately referred to it as 'the greatest British military defeat for many centuries'; the British saved their forces at the expense of the Belgians and the French, and in France 'Dunkirk' was regarded as an act of abandonment and betrayal.[5]

Among the thousands of British soldiers evacuated from Dunkirk was Lieutenant Alan Rook of the Royal Artillery, who published three books of poems during the war, but afterwards gave up literature for the wine trade. Like other young poets at that time, Rook was heavily influenced by the tone and imagery of Eliot's *Murder in the Cathedral*. His 'Dunkirk Pier' opens thus:

> Deeply across the waves of our darkness fear,
> like the silent octopus, feeling, groping, clear
> as a star's reflection, nervous and cold as a bird,
> tells us that pain, tells us that death is near.

Paul Fussell has remarked that Rook's lines show how in the Second World War fear was openly acknowledged as a central part of the experience of battle, in contrast to earlier wars, when it was dismissed as cowardice.[6] The poem's cloudy introspection conveys little of the physical reality of Rook's experience. Yet there was another side to his writing. In a prose journal which he kept during the summer of 1940, after his return from Dunkirk, he makes precise observations of what was going on around him. Rook spoke German and was sometimes called on to interrogate prisoners. His account of an injured German airman shows his capacity for direct reporting:

The prisoner was in a bed with screens around it, in a ward containing about a dozen of our own fellows, all very interested in my arrival. There were cages over both his legs. His head was bandaged. He had a rubber tube coming from his nose with a clip on it, and another tube coming, as far as I could judge, from the bandages round his throat. He was heavily doped and the Sister was doubtful whether she could rouse him.

It seemed very cruel to drag him back from his sleep, but the Sister was most efficient. When he showed signs of consciousness I leaned over and said, 'Good evening, my friend' in German.

The words penetrated through the morphia, and as he opened his eyes there was in them such a blaze of hope and wonder as I never wish to see again. It faded at once when he saw my uniform and he answered my questions dully, only half comprehending . . .

Before we left the hospital, Obergefreiter Hans Schmidt, aged nineteen, of Bonn was dead.[7]

Several books published in 1940 reflected the transition from peace to war. One was W. H. Auden's collection of poetry, *Another Time*. The departure for America early in 1939 of Auden and his friend and collaborator Christopher Isherwood provoked heated debate in literary and intellectual circles, and a question was asked about it in Parliament. Waugh satirized the fuss in *Put Out More Flags*, where 'Parsnip' and 'Pimpernell' are two controversial absentee writers. Louis MacNeice, who had spent most of 1940 in America, returned to England towards the end of the year but defended Auden's decision to stay away. The debate had its serious aspects, concerning the relation of the artist to society, and how far he had responsibilities to it, rather than to his art (*Horizon* had no doubt that art came first). *Another Time* contains some vintage Auden, notably 'September 1, 1939', and the poems which commemorate the deaths in 1939 of two great men, Yeats and Freud. It includes poems written on both sides of the Atlantic, in peace and war, and represents a watershed between the Auden of the 1930s, the political physician diagnosing the sickness of England in the terse colloquial idiom that became instantly recognizable (and widely imitated) as the 'Audenesque', and the reflective religious poet of the American years.

One of the outstanding novels of 1940 was Graham Greene's *The Power and the Glory*. In the 1930s Greene had something in common with Auden; both looked with concern and fascination at the disintegrating world around them.[8] In his earlier fiction Greene was preoccupied with themes of betrayal and pursuit in English urban and suburban settings— the dismal world that critics soon dubbed 'Greeneland'. The flatness of his descriptive prose was enlivened by extravagant metaphor, reflecting, as Greene later confessed, his taste for the conceits of Metaphysical poetry; his fiction also drew heavily on the mass-cultural modes of the thriller and the cinema. The best of his prewar novels, *Brighton Rock* (1938), has all these characteristics, whilst introducing a new element, Catholicism. (Greene had become a Catholic in 1926, but his religion had not previously entered his fiction.)

In *The Power and the Glory* the scene shifts from England to Mexico, about which Greene had already written a travel book, *The Lawless Roads* (1939). The central figure is the unnamed 'whisky priest', who is deeply conscious of his unworthiness, for he is an alcoholic and has broken his priestly vows by fathering a child. The novel is set at a time when there is savage persecution of religion in Mexico, and it focuses on the efforts of the priest to continue as best he can with his sacramental work, even though his life is in danger. *The Power and the Glory* is the most Catholic of Greene's novels and the most accessible of his 'Catholic' novels. Its basic themes—pursuit, suffering, betrayal, the clash of innocence and experience—recur from his earlier fiction, and they have a broad human appeal beyond the specifically Catholic dimension. Greene's own intention was, no doubt, to show how closely the saint and the sinner are related, and the novel is a moving demonstration of the way in which spiritual strength can arise from physical weakness. But it is rather too much of a demonstration; the principal limitation of the novel is that it is too diagrammatic. This is apparent in the opposition between the priest and his principal persecutor, the police lieutenant, an atheist and a decent man, moved by a strong desire to improve the lot of humanity; his ideal is progress as the priest's is salvation. This typological opposition recalls Greene's admiration for the sharp polarities of the morality play. In its most general sense, though, the novel acts out an isolated victim's attempt to preserve his integrity under the pressures both of his own weakness and of external persecution. In the context of 1940 the struggle of individual values against an oppressive political system was an urgent and timely subject.

This theme was also dominant in another important novel of that year, Arthur Koestler's *Darkness at Noon*. Its title invokes Milton's Samson—'O dark, dark, amidst the blaze of noon'—as Eliot had done earlier in the year in *East Coker*. *Darkness at Noon* was originally written in German, but it was first published and widely read in English translation; by the time it appeared Koestler had switched to writing in English, and he later acquired British nationality. He was a Hungarian Jew who spoke several languages and had travelled widely in Europe; he had been a Zionist activist in Palestine and a Communist journalist in Russia. During the Spanish Civil War he was a reporter for an English Liberal newspaper and, secretly, a Soviet agent. He was captured and spent several months under sentence of death in one of Franco's jails before being released. When the Second World War started Koestler was living in France; he was imprisoned as an undesirable alien by the Third Republic before the Fall of France, and

by the Vichy authorities after it. In the summer of 1940 he
escaped to Portugal, with the intention of getting to Britain.
But the British authorities were slow to grant him a visa, so he
travelled to Britain without one. For that he was imprisoned
once more, as he had anticipated. When *Darkness at Noon*
appeared Koestler was in a cell in Pentonville prison, and
London was being bombed; nevertheless, he believed that
any political prisoner in Europe would have happily changed
places with him. He was shortly released and served for a time
in the Pioneer Corps of the British Army before settling down
as an English writer.

Darkness at Noon is set in Moscow at the time of the great
show trials of the late 1930s, when so many of the original
founders of the Russian Revolution were accused of treason
by Stalin and liquidated. It is a political novel of a kind rare in
England, and reflects Koestler's bitter disillusionment with
Communism following his experiences in Spain. Orwell, who
became a close friend of Koestler's, wrote that no native
Englishman could have written *Darkness at Noon*, 'because
there is almost no English writer to whom it has happened to
see totalitarianism from the inside. In Europe, during the past
decade and more, things have been happening to middle-class
people which in England do not even happen to the working class'.[9] Orwell rightly described *Darkness at Noon* as a
masterpiece, and summarized it in these words:

Darkness at Noon describes the imprisonment and death of an Old
Bolshevik, Rubashov, who first denies and ultimately confesses to
crimes which he is well aware he has not committed. The grown-
upness, the lack of surprise or denunciation, the pity and irony with
which the story is told, show the advantage when one is handling a
theme of this kind, of being a European. The book reaches the
stature of tragedy, whereas an English or American writer could at
most have made it into a polemical tract. Koestler has digested his
material and can treat it on the aesthetic level.[10]

The habitually severe Q. D. Leavis also admired Koestler's
novel, and compared him with Conrad: 'Mr Koestler, with
similar creative gifts as a novelist, is also an intellectual with

moral, ethical and sociological preoccupations.'[11] Koestler
was certainly an intellectual with many talents, but he was
not primarily a novelist. In *Darkness at Noon*, though, the
pressure of personal experience and intellectual passion combined with high literary intelligence to produce major fiction.

Darkness at Noon and *The Power and the Glory* both
present isolated and persecuted heroes, striving to maintain
personal integrity in the face of persecution. The two novels
are informed by the clash of ideas in an extreme situation,
sometimes in a diagrammatic way, evident in the debates
between Greene's priest and police lieutenant, and Koestler's
Rubashov and his interrogator Gletkin. The priest and
Rubashov are both dead at the end of their stories; but the
one has achieved a kind of sanctity, while the other has
willingly subordinated himself to the system that destroys him.
Greene was more naturally a novelist than Koestler, but his
novel represents a further development of preoccupations
already rooted in his imagination. Koestler took what he
needed from Dostoevsky and Conrad, but his writing arises
directly from his own discoveries of the terrors of political
ideology.

In the summer of 1940 Britain faced invasion. Inspired by
Churchill, most people took the prospect calmly. Concrete
pillboxes were erected by roads and railways—some of them
are still visible—and thousands of men joined the Local
Defence Volunteers (later the Home Guard) who at first had
little in the way of weapons and equipment. Orwell was one of
them—his health was too poor for the regular armed forces—
and his diary is full of caustic reflections about the poor quality
of national leadership, and speculations about what was really
happening in the war, based on the assumption that in wartime
the truth is frequently suppressed. But even Orwell did not
fully realize the desperate state of the country after the Fall of
France. It was taken for granted by the Germans, the defeated
French, and the neutrals that Britain would seek a negotiated
peace, and Hitler made an offer of terms. Churchill fiercely
rejected any such prospect and the nation was behind him,
though some members of the governing classes were inclined

to take the idea seriously. In retrospect it seems that the magnificent myth of Britain standing alone and saving civilization represented the reality, and that Churchill was right to fight on. But it was, in the Duke of Wellington's words, 'a damned close-run thing'. Defeat might be avoided, with the aid of the physical barrier of the Channel, but victory was a remote, barely conceivable ideal. Churchill's only course was to hold on and hope for American assistance. In the end it came, but very slowly, and was accompanied by conditions which meant the end of Britain's economic independence.

When it became clear that Britain would not make peace the Germans launched air attacks, directed at the RAF and its aerodromes, with the aim of destroying the air defences as a prelude to invasion. By the end of August what became known as the Battle of Britain was under way. On the first anniversary of the outbreak of war, Richard Hillary, now a fighter pilot, took off to engage the enemy:

September 3 dawned dark and overcast, with a slight breeze ruffling the waters of the Estuary. Hornchurch aerodrome, twelve miles east of London, wore its usual morning pallor of yellow fog, lending an added air of grimness to the dimly silhouetted Spitfires around the boundary. From time to time a balloon would poke its head grotesquely through the mist as though looking for possible victims before falling back like some tired monster.[12]

Later that day Hillary was shot down, and was rescued from the sea, badly burned.

A few days later, in what was a tactical error, since the RAF was now hard pressed, the Germans shifted their attacks from airfields to the capital, and the heavy bombing of London began. It was watched by the two young poets who had been together on the first day of war, Roy Fuller and Julian Symons. The latter wrote:

September 7, 1940. Armageddon Day.

On this Saturday afternoon Roy Fuller and I went swimming in Brockwell Park. From a hill in the park we watched the tiny birds far overhead in the blue sky, dozens of them moving in formation undisturbed by the coughing tubes that puffed smoke around them.

Very soon the crump of bombs sounded from the East End, and more flights of birds swam across the sky. The destruction of London had begun. It was a fulfilment of prophecies we had been making for years, and there was nothing to be done about it. We went back to the house in Denmark Hill where I was living, and from there just a few steps up the road to a pub called The Fox Under the Hill, where we drank and played bar skittles until closing time.[13]

Fuller has set down his own memory: 'he and I watched from the height of Brockwell Park, SE24, neat formations of German bombers, with attendant frisky-puppy fighters, moving overhead to raid the London docks. Soon, slanting cloud-mountains of smoke were seen rising from that area. This was more like *Things to Come*.'[14]

The bombing of London had been expected for several years, and was foreshadowed not just in the film *Things to Come* but in newsreels of the bombing of Madrid and Barcelona. The idea haunted the literary imagination. In 1936, the gloomy hero of Orwell's novel, *Keep the Aspidistra Flying*, looked forward with apocalyptic relish to the bombing of London, and in the same year David Gascoyne in his journal recorded standing on Hampstead Heath, thinking, 'What a heart-shaking spectacle it will be from this height some night soon to come, when the enemy squadrons blackening the skies rain down destroying fire upon these roofs.' (The language is biblical, as is so often the case when English writers are in the grip of strong emotion.) In Graham Greene's *A Gun for Sale* (1936) there is a mock air-raid, and in Orwell's *Coming Up for Air* (1939), published not long before the outbreak of war, an RAF plane accidently drops a bomb on an English village. Early in 1941 Orwell wrote, 'On the day in September when the Germans broke through and set the docks on fire, I think few people can have watched those enormous fires without feeling that this was the end of an epoch.'[15] But, he added, life got back to a form of normality remarkably quickly. Connolly expressed a similar sentiment more pithily: 'Most of us who saw London lit up on the night of September the seventh by the fires of the blazing river knew that our world was being ripped up like an old sofa.'[16]

After the initial daylight attack on London the Germans were forced by RAF fighters to switch to night bombing. What became known as the 'blitz' got under way, and apocalypse became routine. The raids continued, sometimes nightly, until the spring of 1941. After Louis MacNeice returned from the United States in December 1940 he volunteered for the navy but was rejected for medical reasons, and he served as a firewatcher during the blitz. Writing about it in an American magazine, he conveyed a curious exhilaration and aesthetic satisfaction, applying a poet's eye that had elements of a painter's:

The one fire which I saw close at hand was very beautiful—a large shop-building that seemed to be merely a façade of windows and these windows were filled to the brim with a continuous yellow flame, uniform as a liquid but bubbling a little at the top of the windows like aereated tanks in an aquarium. There were no wild tongues of flame, no reds or blues or flame-colour proper, but above the building the smoke rolled up and outwards in great soft tawny clouds.[17]

This description of the fire recalls the precise, Ruskinian accounts of natural phenomena in the notebooks of Gerard Manley Hopkins.

MacNeice also described for his American readers a major feature of London life during the blitz: the thousands of people, mostly from poor homes, who slept at night in Underground stations. Henry Moore made a striking series of drawings of them. On the night of 16–17 April 1941 there occurred the heaviest German raid so far, which caused much damage and loss of life. MacNeice wrote of it:

When the All Clear went I began a tour of London, half appalled and half enlivened by this fantasy of destruction. For it was—if I am to be candid—enlivening. People's deaths were another matter—I assumed they must have been many—but as for the damage to buildings I could not help—at moments—regarding it as a spectacle, something on a scale which I had never come across.[18]

At the same time, MacNeice was impressed with the speed with which emergency repairs were carried out and the streets cleared up and made passable within a few hours: 'It was all the difference between a raw wound and a wound that has been dressed.'

Graham Greene also found a curious satisfaction in the danger and destruction. Like other writers who had come to maturity in the 1930s he had for some years been expecting the violent collapse of a rotten civilization, and now it seemed to be actually happening. In Greene's case this shared perception was reinforced by the romantic primitivism of his personal mythology. He wrote of feeling at home in London or other bombed cities,

because life there is what it ought to be. If a cracked cup is put in boiling water it breaks, and an old dog-toothed civilization is breaking now. The nightly routine of sirens, barrage, the probing raider, the unmistakeable engine ('Where are you? Where are you? Where are you?'), the bomb-bursts moving nearer and then moving away, hold one like a love-charm.[19]

Greene served as an air raid warden, and was on duty throughout the night of 16 April, engaged in extricating the dead and injured from bombed buildings. He set down his experiences in 'The Great Blitz of Wednesday, 16th April'.[20] The following morning Herbert Read walked across London from the Bank to Piccadilly, having to follow a circuitous route through the glass-strewn streets. Near St Paul's the roads were closed, so he crossed the river by Southwark Bridge. He crossed back by Blackfriars Bridge only to find that the Strand was blocked, so he continued on his way via Covent Garden, 'past burning ruins in Leicester Square, and so into Piccadilly, looking sultry under a smoke-screened sky'.[21] Perhaps the best brief comment on the bombing comes in the arresting opening sentence of Orwell's pamphlet, *The Lion and the Unicorn*, which brings into focus the juxtaposition of barbarism and high technology that characterises modern war: 'As I write, highly civilized human beings are flying overhead, trying to kill me.'[22]

Writers on an Island

I

Britain's geographical situation has long been an uncertain blessing: to be an island may be a rather fine thing, but to be insular is often regarded, especially by intellectuals, as a fault. During the war, John of Gaunt's speech from *Richard II* about the virtues of the 'sceptered isle' was frequently invoked:

> This happy breed of men, this little world,
> This precious stone set in the silver sea,
> Which serves it in the office of a wall,
> Or as a moat defensive to a house . . .

In 1940 the patriotic rhetoric applied literally; the German army which had swept across Western Europe was halted by the moat of the Channel. Geography, combined with British air power, saved the nation from invasion, though no one knew for how long. Half-buried historical memories of this circumstance may have played a part in British resistance to schemes of European integration during the next half-century.

For middle-class writers accustomed to foreign travel, the severance from Continental Europe was particularly trying. Connolly compensated as best he could by publishing in *Horizon* many articles on European literature and art, and by seeking contributions from Continental writers in exile. France, above all, inspired sentiments of nostalgia and longing, and in 1943 Arthur Koestler ironically diagnosed a condition he called 'French 'Flu':

The people who administer literature in this country—literary editors, critics, essayists: the managerial class on Parnassus—have lately been affected by a new outbreak of that recurrent epidemic, the

French 'Flu. Its symptoms are that the patient, ordinarily a balanced, cautious, sceptical man, is lured into unconditional surrender of his critical faculties when a line of French poetry or prose falls under his eyes . . . For the frustrated lover of France even the names of Paris underground stations (Vavin, Les Buttes Chaumont, Réaumur-Sevastopol, Porte des Lilas) become the nostalgia-imbued stimuli of conditioned reflexes: first there is the flutter and twitch of the heart, then the mucus of the French 'Flu begins to flow.[1]

Meanwhile, throughout German-occupied Europe Britain had acquired a new significance, particularly because of the broadcasts of the BBC, a symbol of hope and resistance. The mysterious, foggy island of Continental mythology became a beacon of freedom. And although the war had cut Britain off from Europe, there was a sense in which Europe had come to Britain, in the shape of exiled foreign governments and their armed forces. Wartime London became unprecedently cosmopolitan, the streets filled with unfamiliar uniforms and polyglottal speech. After the liberation of Paris in August 1944 English writers made enormous efforts to get back there. Philip Toynbee was one of those who succeeded, and, flushed with French 'Flu, wrote an eager article in *Horizon* describing his discoveries on the Parisian literary scene, in the gorged manner of a boy raiding a sweetshop. In a sour letter to the magazine defending English wartime writing—which Toynbee had used the occasion to knock—the critic and editor John Lehmann remarked that Toynbee could not possibly have read all the books he referred to. Elsewhere British soldiers were breaking down insular barriers and making their own cultural discoveries; in Italy many of them acquired a taste for wine and opera.

Darkness, loneliness and displacement were only one aspect of the wartime condition. The mixing of social classes and the physical mobility brought by military service changed attitudes and enlarged horizons. Isolation and the threat of violent death were to some extent balanced by a new sense of community and shared hopes for the future. During the war years the collective mood moved to the Left. There was a determination that the postwar future would be better than the recent past of industrial depression and mass unemployment. The American vice-president, Henry Wallace, described the twentieth century as the Century of the Common Man. The 'Beveridge Report' of 1942 laid the foundations for a comprehensive welfare state, and was eagerly seized on as a charter for the future. In retrospect, the widespread passion for social betterment looks naïve, given that Britain was largely bankrupted by the war, and that her economic prospects after it were grim. Indeed, the right-wing historian Corelli Barnett has argued that the postwar insistence on full employment and the welfare state were major factors in Britain's economic decline, by diverting resources from the essential task of rebuilding the industrial base and re-establishing lost overseas markets. At the time such arguments were heard from some Tories—though by no means all—but they were swept away in the prevailing tide of egalitarianism and radical enthusiasm. Party politics were frozen during the war; the Labour Party had joined Churchill's coalition in 1940, and the established parties agreed not to oppose each other at by-elections. But the Common Wealth Party, an idealistic neo-socialist group that did not observe the truce, had electoral successes and ended the war with a number of MPs.

Arthur Koestler, a sympathetic but sharp observer of English life and mores, described a representative type of young serviceman whom he had often come across while lecturing to army audiences. He defined him as the 'thoughtful corporal', who is serious-minded and intellectually aspiring, though not highly educated:

I imagined seeing you, on Saturday afternoons, scanning the dismal Y.M.C.A. library, or asking the lymphatic girl at the W. H. Smith bookstall for a Penguin which is out of print . . . You feel vaguely attracted by Common Wealth . . . you keep a diary at irregular intervals, plan to write a short story on army life for *Tribune* or *New Writing* . . . You see your future in a rather grim perspective of W.E.A. evening classes, night reading, and an elaborate saving plan for a week's visit to France.[2]

Though based on sound observation, Koestler's picture is somewhat distorted by its literary antecedents, such as E. M.

Forster's Leonard Bast and, perhaps, Hardy's Jude. In 1944, when Koestler wrote, this young man may well have found that the name of the Labour Party tasted 'like stale beer'. But it was those like him, in their millions, who gave Labour its great victory the following year. And in the postwar era he probably found a brighter future and more opportunities than Koestler envisaged.

The 'thoughtful corporal' was particularly representative in his constant search for something to read. Men and women in uniform, or on other war work, read a great deal in their spare time or the long intervals between periods of action. The solitary reader in an army or air force base or on a ship, often reading against the distraction of others talking in the room or the noise of the radio, was a typical figure. The demand for books was not easily satisfied, as paper was strictly rationed, and 20 million volumes were destroyed by fire in an air raid on the publishing quarter of London in December 1940, a loss which was never made good. There was a desire for long classical novels, which provided sustained and absorbing distraction; Trollope was particularly popular. But books of any quality and on practically any subject were in demand, and publishers, provided they had secured an adequate paper ration (based on the volume of their prewar sales), did very well. Koestler describes his young corporal as reading the *News Chronicle*, a left-Liberal daily, and *Tribune* and the *New Statesman*, which were socialist weeklies. He would probably have also read *Picture Post*, a popular illustrated weekly which provided a pictorial record of events in a pre-television age, and which took a left-wing editorial line, populist and egalitarian, demanding a better world after the war. *Picture Post* and Penguin books were powerful opinion-formers, and so were the army education classes and discussion groups, like those to whom Koestler lectured. These classes had been started in order to keep up morale in the services, but many Conservatives suspected, rightly, that they were seedbeds of left-wing ideas.

The corporal aspired, too, to write something for *Penguin New Writing*. This paperback miscellany was the most popular and widely read of all wartime literary publications. Indeed, it remains one of the most interesting cultural phenomena of the twentieth century. It was edited by John Lehmann, an Old Etonian critic who combined Bloomsbury affinities—he had been assistant to Leonard and Virginia Woolf at the Hogarth Press—with left-wing sympathies and proletarian tastes, sexual as well as ideological. His sisters Rosamund and Beatrix were celebrated as, respectively, novelist and actress. In the 1930s Lehmann had edited an earlier version of *New Writing*, which provided a platform for anti-fascist European writers, and which continued into the war years as an occasional hardback publication, *New Writing and Daylight*. *Penguin New Writing* launched in 1941, was an offshoot in paperback, aimed at a wider audience, and with Penguin Books as backer it was assured of a generous paper ration, based on their rapid expansion in the late 1930s. It published poems, stories, and reportage by previously unknown men and women caught up in the war, together with literary material by established writers, and articles on literature, theatre, music, and art. There were reproductions of paintings, and illustrations of theatrical productions and set designs. It was a remarkably successful formula. Because of the ban on new magazines it was described as a 'miscellany', a convenient device which was widely invoked during the war. *Penguin New Writing* came out three or four times a year in impressions of about 75,000 copies. Each issue was likely to have had many readers, and may well have been passed around until it fell to pieces. This indicates an extraordinarily wide circulation for a serious literary and cultural publication. *Penguin New Writing* eschewed overt politics, but it represented the ideals and interests of Koestler's thoughtful corporal, and can be fairly thought of as embodying the unified culture of wartime Britain. As Robert Hewison puts it:

These slim volumes, with pastel covers and fragile wartime paper, became emblems of cultural survival which no Ministry of Information publication or film documentary could manufacture. Tucked in battle dress pocket or gas-mask holder or factory overall, they created a link between the civilian and military worlds, and in the years of rationed poetry and imaginative confinement, it is no surprise that they sold out as soon as they appeared on the bookstalls.[3]

Horizon and *Penguin New Writing* shared the mission of keeping high culture alive in wartime, and had an overlapping readership. But *Horizon* was defiantly élitist, whereas *Penguin New Writing* was unobtrusively populist and looked to a wider audience (though Connolly complained that the circulation of *Horizon* was limited by the restrictions on paper supply and that he could have sold many more copies than he did). *Horizon* did sometimes publish work by unknown writers, and claimed dispiritingly to be always looking for it but seldom finding anything good enough; Lehmann's standards were a little less severe, in the hope of discovering new talent. But the thoughtful corporal and those he stood for might have been baffled by *Horizon*'s assumption that its readers were all at home in French, and would pick up allusions to symbolist poetry or baroque architecture, and they might well have been turned off by Connolly's more attitudinizing editorials. Nevertheless, *Horizon* and *Penguin New Writing* were complementary rather than opposed in their aims and achievements. (They both ceased publication in 1949–50.)

As well as the popularity of reading, there was a great demand for classical music, drama, ballet, and art exhibitions, and an official body was brought into being to cater for it, the Council for the Encouragement of Music and the Arts. After the war it became the Arts Council. At the beginning of the war the paintings had been removed from the National Gallery for safety, but from time to time, by popular demand, a single Old Master was brought back and exhibited; the Gallery was also used for widely attended piano recitals by Myra Hess and other famous performers. Similar activities, modest in themselves but showing a need for something finer than the privations and brutalities of the age, were repeated all over the country. The extent of the interest should not, of course, be exaggerated; a stock type in wartime short stories and sketches is the sensitive, thoughtful, art-loving serviceman who is oppressed by the coarse philistinism of those around him, whether Other Ranks or officers. All the same, the proportion of the population who had a taste for what we have subsequently come to regard as 'high culture' was remarkably high. At that time the taste extended over class barriers and social divisions; the notion that high culture is inherently 'middle-class' or 'bourgeois' and thus ideologically tainted is a much later development.

It is true that the love of art had an element of escapism to it, whether it took the form of reading Trollope or listening to a recital of Beethoven piano sonatas. This may not be the finest way of responding to art but at the time it was understandable. Some kinds of art, where there was an emphasis on sensuous delight and spectacle, catered in very direct ways for the need to escape; ballet was a particular example. Something similar may account for the love of a kind of poetry that was sonorous to the ear, and rich in imagery and language. The ideal of brief escape to a rich world—in more than one sense—cut across the earnest egalitarianism of wartime society. It had long been catered for by Hollywood films—the arrival of American soldiers later in the war gave the celluloid world a kind of substance—while the British cinema provided its own version in the popular costume-dramas made by Gainsborough Studios, starring James Mason and Margaret Lockwood, such as *The Man in Grey* (1943) and *The Wicked Lady* (1945). The glamour of 'high-life' featured in the popular songs that poured continually from radios and which appealed to all classes. In one of them 'there were angels dining at the Ritz | And a nightingale sang in Berkeley Square', and in another a couple—assumed to be married, for the sake of propriety—spent their last night together before the man's departure for the war in 'Room 504' of a grand hotel: 'We hadn't dared to ask the price, That kind of thrill can't happen twice'.

There is a brilliant literary treatment of wartime attitudes and beliefs in Kingsley Amis's novella, 'I Spy Strangers', of which Paul Fussell has written, 'Amis has proposed, in only about 20,000 words, a compressed "war novel" registering the moral meaning of his military experience against a background of European politics and the British social framework.'[4] The European war has ended before it opens, and the army signals unit that it deals with has never been in action. The place is defeated Germany, and the time is the summer of 1945; the results of the British General Election are expected shortly, but meanwhile the war in the Far East drags on, perhaps for a long time to come; the atomic bomb has not yet been dropped on Hiroshima. The signals unit is losing coherence; odd military personnel from elsewhere are attached to it, others are sent elsewhere, and morale is drooping. The men are all eager to get back to England, but some of them at least will have to be sent to the war against Japan. The decision lies with the commanding officer, Major Raleigh, a weak-minded martinet who cannot keep his unit in proper order but who is prepared to act vindictively against men and officers who displease him.

'I Spy Strangers' is a finely observed study of the inter-actions of men in an artificial institutional environment, after the immediate pressures of war have been lifted. It was first published in Amis's collection of stories, *My Enemy's Enemy* in 1962, but was presumably written some years earlier; two other stories dealing with the same unit and some of the same characters were originally published in 1955–6. It is evident from the story that when Amis wrote it his sympathies were firmly on the Left; his swing to the Right in the 1960s was yet to come. The principal opposition is between Major Raleigh, who is bitter about not having been sent to command a signals unit at the Potsdam Conference, and Lieutenant Archer, the young officer who is the hero of the story and the focus of authorial sympathy. Archer is cleverer and better educated and perhaps more cunning than Koestler's thoughtful cor-poral; he was at Oxford before going into the army, like Amis himself, and will be returning after the war to finish his degree. But he could be the corporal's cousin; they have the same loosely left-wing ideals, cultural interests, and hopes for a better future. There is another and very interesting character, Sergeant Doll, an intelligent and articulate rep-resentative of the far Right, who believes that the great coming task, once Germany and Japan are out of the way, is the defeat of Communism. He is presented as a monster, but with a hint of sympathy; the later Amis might have thought Doll's ideas had a lot to be said for them, and dismissed the worthy Archer as a snivelling Lefty.

However, in the framework of the story as Amis wrote it, the fundamental opposition—in both values and dramatic conflict—is between Raleigh, who is a Conservative out of opportunism rather than ideological conviction (and despised by Doll), and Archer, the Labour-voting new man. The thoughts and utterances that Amis gives them have a representative quality beyond their specific fictional context. Raleigh cannot believe that Labour stands any chance in the election, but still has a sense that things may not go his way: 'And yet the major was uneasy. Something monstrous and indefinable was growing in strength, something hostile to his accent and taste in clothes and modest directorship and ambitions for his sons and redbrick house at Purley with its back-garden tennis-court'. When the election results come out Raleigh is shattered, particularly when he discovers that the Labour candidate in his own seemingly secure Tory constituency has won with a thumping majority. Archer on the other hand is quietly elated, and when invited by Sergeant Doll to drink a toast 'to England' he thinks that he will not drink to Doll's England, nor to that of his brother officers, but to the England of some of the Other Ranks whom he admires, 'and absolutely my England, full of girls and drinks and jazz and books and decent houses and decent jobs and being your own boss'. This was certainly the ideal of a great many of the young men who had given up several years of their lives to the war. But it is a private, mildly hedonistic ideal, not altogether congruous with its public occasion, the triumphant election of a socialist government with a reforming programme. The conflict of attitudes in the story is firmly resolved in Archer's favour, but in a personal rather than a political way, when, much to Raleigh's consternation, he is able to leave the army early and escape the Far East war, where Raleigh had tried to send him, in order to resume his studies at Oxford. 'I Spy

Strangers' is an outstanding work of short fiction, sharply reflecting its historical moment.

II

Although reading was so popular in wartime it was not a good period for the novel. No significant new talent emerged between 1939 and 1945, in contrast to, say, the late 1920s, which saw the first novels of Henry Green, Graham Greene, Christopher Isherwood, Rosamund Lehmann, and Evelyn Waugh, or to the 1950s, when Kingsley Amis, William Golding and Iris Murdoch all published their first novels in a single year. During the war there was a sharp fall in the number of new novels published, from 4,222 in 1939 to 1,179 in 1945. The shortage of paper would have been a major factor in this decline; but it also reflected the fact that young writers on active service or engaged in other kinds of war work lacked the freedom and time needed to write novels, and so concentrated on shorter literary forms: poetry (usually the brief lyric), stories, and reportage. Nevertheless, some novels of high quality did appear during the war.

Virginia Woolf committed suicide in 1941, having completed but not revised her last novel, *Between the Acts*; Leonard Woolf prepared the manuscript for publication, and brought it out later in the year. Like her other novels it traces the interplay and movement of fine sensibilities, in a subtle and intricate and frequently poetic prose. It expresses a modernist conviction about the power of art to transcend the flux of existence; at the same time, it is much more firmly located in history than Woolf's earlier work. *To the Lighthouse* (1927), it is true, had introduced the First World War in an extended parenthesis, but the novel as a whole had aspired to the timelessness of the self-contained aesthetic entity, symbolized by Lily Briscoe's painting. *Between the Acts* echoes in its title Eliot's *l'entre deux guerres*, and it is set at the very end of the inter-war period, in an archetypal English village and its country house, Pointz Hall, on a single day in June 1939. War threatens, and Woolf introduces the menacing, low-flying aeroplanes that recur in the writing of the late 1930s. There

are more symbolic representations of impending violence, as
in the strange discovery of a snake choked by a toad it is trying
to swallow. The past is re-enacted in the pageant of English history devised by Miss LaTrobe; she is in her own way an artist,
but is at the same time a rather absurd figure, whom Walter
Allen has suggested is Woolf's deliberate burlesque of herself
as an artist. The pageant is a mode of art, but at that point in
history it is also an elegiac celebration of a form of life that
may be on the verge of dissolution. At the same time, the
novel reaches back beyond recorded history to prehistory, in
the consciousness of the elderly widow Mrs Swithin, whose
favourite reading is the early chapters of H. G. Wells's *Outline
of History*. Despite its unrevised state, and Woolf's dissatisfac-
tion with it at the time of her death, *Between the Acts* is
arguably her best novel; certainly, it is her most approachable.

Patrick Hamilton's *Hangover Square* (1941), is, compared
with Woolf's exploratory modernism, a novel of traditional
physical and psychological realism. Hamilton carefully renders
a certain kind of seedy London life, focusing on the pubs and
furnished flats of Earls Court, with an occasional expedition to
London's seaside outpost at Brighton. But *Hangover Square*
shares with *Between the Acts* a strong evocation of the last
days of the doomed peace. The story moves through the
months of 1939, and follows the mental collapse of the amiable but ineffective drifter, George Harvey Bone; Hamilton
impressively sustains the parallels between personal and public
disintegration. They converge on the day war breaks out,
when, at the climax of the novel, George murders his odious
girl-friend Netta and her fascist lover, against the background
of Chamberlain's broadcast and the first air raid warning.
Hangover Square is both a masterly study of obsession and a
work of precisely observed historical fiction. These novels by
Woolf and Hamilton document, in their very different ways,
the sense of the ending of an age.

A third novel from 1941, Rex Warner's *The Aerodrome*, is a
work of allegorical fiction, not intended to be realistic. At the
same time, it attempts to engage with the contemporary world,
and to show the nature of fascism. The story begins with an

old-fashioned English village, furnished with vicar and squire and cheerful boozing yokels, drawn with comic-strip simplicity. The familiar easy tenor of life is brutally disturbed when a new aerodrome is built nearby and the 'Air Force', which seems more like the Waffen SS than the RAF, takes over and transforms the life of the neighbourhood. Warner evidently planned and started the novel before the war, and had to insert a note saying that of course the Air Force in question was not that of his own country. The commander of the aerodrome is the Air Vice Marshal, an articulate fascist intellectual who speaks of the need to introduce efficiency and order and 'to make the world clean'. John Lehmann wrote admiringly of *The Aerodrome*, comparing it with *Between the Acts*, seeing both books as profound parables set in a version of English village life, with poetic symbols played against surface realism. This is broadly true of *Between the Acts*, but hardly, I think, of *The Aerodrome*, where the allegory is often crude and forced. Insofar as it retains a thematic interest, it works not so much as an exposure of fascism as a parable of the destruction of a traditional rural order by advancing modernity. *The Aerodrome* caused Warner to be described as an English Kafka, but that comparison now looks wide of the mark. The work's real antecedents lie in the schoolboy jokiness of the early Auden circle, of which Warner was a part; it recalls the solemn authorial tone and the taste for random violence and melodramatic coincidence in Edward Upward's 'Mortmere' fantasy 'The Railway Accident', and the bizarre evocation of village life in the Auden-Isherwood play, *The Dog Beneath the Skin*.

The bombing of London in 1940–1 is prominent in three novels published in 1943: Graham Greene's *The Ministry of Fear*, James Hanley's *No Directions*, and Henry Green's *Caught*. Greene's novel was originally published as an 'entertainment', like his prewar thrillers, though the distinction between his entertainments and his serious fiction was always arbitrary and he eventually dropped it. In books such as *It's a Battlefield* and *The Confidential Agent* Greene had presented London with an intense cinematic responsiveness to the life of the streets; in fact, in 1938 he told Julian Maclaren-Ross that as far as possible he liked to restrict himself to London settings.[5] This was a restriction that Greene shortly overturned, for from 1940 onwards most of his novels were set out of England, beginning with *The Power and the Glory*. His next novel, *The Ministry of Fear* is at its best as an evocation of London during the air raids; there are some fine descriptive passages, based on Greene's journal of the blitz. In this novel the realistic cinema that had influenced his fiction in the Thirties gives way to something closer to the stylized distortions of expressionism, and it may have been this that induced Fritz Lang to make a film of it. Considered as a thriller, though, *The Ministry of Fear* is confused and confusing. The central figure, Arthur Rowe, is one of Greene's familiar lonely and soon-to-be-hunted characters. He has been acquitted of murdering his invalid wife in a mercy-killing, but before long he gets caught up in a network of fifth-columnists and Nazi agents, and finds himself on the run, pursued for a murder he has not committed. For much of the latter part of the novel, Rowe has lost his memory, which adds to the general air of incoherence. It moves finally, and unusually for Greene, to a kind of happy ending. At one point, Rowe, making his way through the ruins of Holborn, looks back to the last day of peace, in the retrospective movement that characterizes much wartime fiction: 'Peace had come to an end quite suddenly on August the thirty-first.'

James Hanley's *No Directions* is a short and concentrated novel, heavily influenced by Joyce's *Ulysses* in its reliance on stream-of-consciousness and interior monologue, and its observance of a tight unity of time, space, and action. The scene is a large house in or near Chelsea which has been converted into small flats, and the action is confined to a single night of the blitz. This formula brings together a handful of representative Londoners, first in their separate rooms, and then sheltering together in the cellar: an air raid warden and his young wife, an elderly couple, an airman on leave with his family, a painter and his wife (who are as much concerned with protecting his paintings as themselves), and two people who have wandered in from outside: a drunken sailor, and a woman who was once the painter's model and who has come to see him, but is turned away by his wife. Hanley juxtaposes and interweaves these separate lives against the crash of bombs and anti-aircraft guns. The use of stream-of-consciousness and a total abstention from authorial comment means that there are elements of opacity, but, as the title implies, that may be a characteristic of the situation the characters are in. *No Directions* is a compelling small-scale work, that reads like the miniaturized version of a much longer novel.

Henry Green's *Caught* is an outstanding novel about the early phase of the war, just as his *Party Going*, published on the eve of war in 1939, had caught superbly the state of feeling of the last days of peace, in an account of a group of wealthy tourists prevented from travelling abroad by the descent of fog on a main-line London station. Green, who was of the generation of Graham Greene, Christopher Isherwood, George Orwell, Anthony Powell, and Evelyn Waugh, and had known some of them at school and university, never achieved a comparable reputation. He has long been admired by critics and other novelists, but neglected by the general reading public. This may be because Green, without being obviously modernist or 'experimental', is disconcertingly original and sometimes difficult. Whereas other novelists give the impression of cutting patterns out of good existing cloth, Green seems to begin by weaving the cloth. This much is apparent in his style, which is like no one else's; syntax is distorted and elided, modes of speech interwoven or superimposed; time is treated fluidly, and strange metaphors introduced. The result can seem highly mannered, but everything is done in the interests of what Green sees as greater precision and intensity of perception.

Caught draws on his own experience as an auxiliary fireman in London during the first year of the war. Like Waugh's *Put Out More Flags*, the novel begins in September 1939 with the advent of the blackout, then moves through the empty months of Phoney War, to the arrival of real war in the Norwegian campaign and at Dunkirk; it culminates in the great fire raids

on the docks in the autumn of 1940. The central figure is
an auxiliary fireman, Richard Roe, a rather bored, well-off
widower with a small son (being looked after in the country by
his sister-in-law), who takes a while to adjust to the working-
class culture of the fire station. The other main character
is Roe's boss, Albert Pye, a regular fireman who has been
promoted to take charge of the fire station, though he is not
very efficient, and is barely able to cope with the job. The
encounters between Roe and Pye reflect the amiable class-
collisions that are such a common feature of English fiction.
There is an element of basic realism in *Caught*, as Green
shows the lives of the firemen during the Phoney War, divided
between training, boozing, and looking for girls. Green
catches the speech and general quality of working-class life as
effectively here as he had done in his second novel *Living*
(1929), and much of the book is very funny. A more sombre
aspect emerges, though. Some time before the war Roe's son
had been abducted from a toyshop, then recovered unharmed
after some hours. Pye is tormented by the knowledge that the
woman who took the boy is his mentally disturbed sister;
there is further torment for him in the memory, which he
desperately tries to suppress, of his sexual assault on his sister
in their youth, an experience that may have led to her later
disturbance.

  Green's stylistic devices convey an often startling juxtaposition of inner and outer, personal and public. Here, for
instance, is his account of Roe's arrival at the fire station at the
outbreak of war, where an inert rendering of the external
world is strangely transformed by his memory of his dead
wife's eyes:

  For several hours they had all to stand in different groups only
to be regrouped, creatures of the utter confusion the London Fire
Brigade creates.
  They were mute in a vast asphalted space. The store towered
above, pile after dark pile which, gradually, light after light went
darker than the night that was falling and which he dreaded. For
twenty minutes at dusk the scene was his wife's eyes, wet with tears
he thought, her long lashes those black railings, everywhere wet, but,

in the air, the menace of what was yet to be experienced, the beginning.

Earlier the balloons had been a colour of the blade of a knife.

Green's writing can provide a virtuoso display of clashing linguistic registers. He is given to breaking the characters' interior monologues with a sudden intervention of authorial comment, as in this proleptic instance:

Months afterwards, when the blitz began, flame came to be called 'a light', they talked of 'putting the light out' instead of 'getting the flames down'. But on those first evenings there was not one Auxiliary, fresh to the black-out, who could foresee the white flicker, then the red glow which spread and, close to, the greedy extravagance of fire which would be bombed and bombed and bombed again to increase the moth's suicide it was for firemen.

Eventually the blitz comes, and Richard acts with courage and professionalism in fire-fighting at the docks, breaking out of the neurotic self-absorption that he has been caught in for most of the narrative. Green holds back until the last few pages of the novel, and then gives a superb account of the blitz, conveyed in three counterpointed discourses: Richard's description of his experiences to his not-fully comprehending sister-in-law ('there was nothing in what he had spoken to catch her imagination'); his interior monologue; and parenthetical authorial commentary, which corrects or modifies what Richard is trying to recall:

'Well, when we got round those buildings I told you about, they were great open sheds really, for keeping the weather off the more expensive timber, we were right on top of the blaze. It was acres of timber storage alight about two hundred yards in front, out in the open, like a huge wood fire on a flat hearth, only a thousand times bigger.'

(It had not been like that at all. What he had seen was a broken, torn-up dark mosaic aglow with rose where square after square of timber had been burned down to embers, while beyond the distant yellow flames toyed joyfully with the next black stacks which softly merged into the pink of that night.)

The authorial interventions point up the strange and terrible beauty of the fires, which had so fascinated MacNeice. We see that when the blitz comes the firemen, who had such an easy time during the Phoney War, are in just as much danger as front-line soldiers, as indeed are the civilian population at large.

In 1943 Arthur Koestler published *Arrival and Departure*, his first novel to be written in English, which draws on his experiences in 1940 when he was fleeing Nazi-occupied Europe. It shows how quickly Koestler had learnt to write effective English prose, but it has none of the power and historical insight of *Darkness at Noon* and deserves no more than a passing mention. It is, though, of interest in providing a very different perspective on the war from the work of the unavoidably insular writers in Britain.

At one point Koestler's hero describes how as a political prisoner in his Central European country, he had been put on a train which contained several wagons containing gypsies destined for sterilization, healthy Jews intended for forced labour, and old and sickly Jews who are going to their deaths. After a long and horrific journey he observes the gassing of the Jews in vans by the side of the railway line. Then it is discovered that he has been put on the train by a mistake, and he is returned to his prison cell. Koestler based this episode on accounts by a Polish eyewitness of the early stages of the Nazi Final Solution. When it was published separately in *Horizon* some readers accused him of exaggeration and atrocity-mongering, and even of sheer invention. He found such responses indicative of the physical and imaginative gulf that separated the British from the day-to-day realities of life in Continental Europe. English fiction and memoirs of the war years are full of long and uncomfortable journeys in crowded, dimly lit trains. But all over Europe trains were transporting people in infinitely worse conditions to unspeakable destinations.

III

One of the most admired writers of serious literary fiction was the Anglo-Irishman Joyce Cary, who came late to literature;

he was born in 1888 and did not publish his first novel until 1932. Cary may have been a little overrated in the 1940s, but he deserves something better than his present neglect. His trilogy comprising *Herself Surprised*, *To be a Pilgrim*, and *The Horse's Mouth* appeared between 1941 and 1944. These novels present the linked fortunes of three characters: first, Sara Monday, whom we first meet as a young servant girl at the end of the last century; she marries a rich man, lives like a lady for a while, then falls from grace and fortune; the next book deals with Mr Wilcher, the Nonconformist bachelor lawyer for whom she works as cook and housekeeper and provider of sexual solace, and who hopes to marry her; the final volume is the story of Gully Jimson, a poor but artistically ambitious painter, who lives with Sara for a time. Cary presents each of these characters in turn in a first-person narrative, without authorial comment, and the result is a remarkable feat of impersonation, of the rendering of character from within, in all its plenitude.

The Horse's Mouth was a huge success in 1944. Apart from its power as a narrative, the novel is a defiant assertion of the transcendent value of art, which made it popular at a time of so much destruction and loss. *The Horse's Mouth* is, strikingly, a painter's book. Its visual responsiveness and awareness of the technical problems and possibilities of planning and executing a work of art reflect Cary's own early training as a painter. Unlike Mr Wilcher, Gully Jimson is not interested in historical change, or in anything other than staying alive and planning and getting the materials for his next painting. But the year is 1939; the novel ends in an extraordinary epiphany in the first days of the war, with Gully and some eager young assistants painting an enormous mural on the inner wall of a condemned chapel, even as a team of workers from the local authority are demolishing the building as unsafe. The symbolism is assertive; civilization may be destroyed but the spirit that produces art endures in spite of everything.

One can see why *The Horse's Mouth* went down so well, and it is still a vigorous and readable narrative, dominated by the large personality of Gully Jimson, the kind of book that reviewers like to call 'rumbustious'. All the same, Gully is a great bore, with his driving egoism and button-holing tone, and his invocations of Blake and Spinoza. I sense a deep sentimentality in Cary's presentation of Gully that is absent elsewhere in the sequence—even where one might expect it, in the rendering of Sara. Gully embodies some crude popular myths: the loveable, cheerfully amoral bohemian rogue; and the great artist whose behaviour is justified because he is an artist, and who is thus entitled to cheat tradesmen, and exploit his women and beat them up at regular intervals. I agree with Walter Allen that the far subtler *To Be a Pilgrim* is Cary's major work in the trilogy: 'the creation of Wilcher in *To Be a Pilgrim* is the greater feat. Wilcher is a true original: there is no other character like him in the range of our fiction.'[6]

Evelyn Waugh's *Brideshead Revisited*, published just as the European war was ending, is an elaborate work of retrospective fiction, which was a new departure for Waugh and, as he ruefully remarked, 'lost me such esteem as I once enjoyed among my contemporaries'. Readers who appreciated the satirical thrusts and sharp laconic prose of Waugh's earlier fiction were disconcerted by the opulent, leisurely writing, the soaring flights of nostalgia, the fascination with the aristocracy, and the unexpected religious dimension, as Waugh, a Catholic convert in 1930, emerged for the first time as an explicitly Catholic novelist. But if it lost Waugh some readers it gained him many more; *Brideshead Revisited* sold well when it came out, and has continued to do so, while the television serial made in the early 1980s, which extravagantly heightened the glamour and the air of aristocratic decadence, extended its appeal. *Put Out More Flags* had continued the manner and some of the characters of Waugh's early fiction, showing the Bright Young Things, their youth and brightness somewhat dimmed, trying to turn into patriots. That book may suffer from its divided aims, but it contains some of Waugh's best comic writing; it also marks the conclusion of the first phase of his fiction. He had already suggested new approaches in *Work Suspended*, an abandoned fragment of a novel which breaks off with the air raid sirens sounding on the first day of war. In this piece Waugh first adopts the first-person narrator and leisurely, reflective prose that he was to exploit so fully in *Brideshead Revisited*.

The narrator of that work is Captain Charles Ryder, who in peacetime had been an artist who specialized in drawing fine buildings, and as the subtitle indicates the book consists of his 'sacred and profane memories'. The memories begin in Oxford in the early 1920s, when Charles as an undergraduate meets the glamorous aristocrat Sebastian Flyte, and continue as Charles becomes more and more involved with Sebastian's family; his elder sister, Lady Julia; his acerbic and teasing younger sister, Cordelia; and his devout and possessive mother, Lady Marchmain, who comes from an old Catholic family. The settings move between Brideshead Castle, the Marchmains' noble country house, and Oxford, fashionable London, and the smarter parts of 'abroad'. But as Waugh came to realize, the retrospective visions were all rooted in the present reality of war. In his preface to the revised edition of 1960, Waugh remarked of 1944, when he wrote the novel on extended leave from the army:

It was a bleak period of present privation and threatening disaster—the period of soya beans and Basic English—and in consequence the book is infused with a kind of gluttony for food and wine, for the splendours of the recent past, and for rhetorical language, which now with a full stomach I find distasteful ... it is offered to a younger generation of readers as a souvenir of the Second World War rather than of the twenties or of the thirties, with which it ostensibly deals.

Most critics have tended to agree with Waugh's criticisms of the book, and believe that his revisions made only very marginal difference to the shortcomings that he acknowledged. Waugh was attempting something new in his writing, but it was surely unwise of him to abandon so completely the satiric edge and the comic sense that are his principal strengths as a novelist. Much of *Brideshead Revisited* is weak in conception, and indulgent to the point of absurdity in writing and attitudes. There are, of course, redeeming moments and passages, provided by such incisive and cleared-eyed figures as the aesthete Anthony Blanche and the young Cordelia, but not enough of them. A central problem is that the whole story is presented through the consciousness of Charles Ryder, who is not Waugh himself, but is not very distanced from him; one can compare the effect Waugh was later to achieve in *The Ordeal of Gilbert Pinfold*, in which the central figure is clearly very Waugh-like, but is presented with careful detachment. And Ryder is, I think, one of Waugh's least interesting characters: a snob and a sentimentalist, absurdly bedazzled by the Marchmains. His memories may well be distorted by the nostalgic haze in which they are presented; indeed, *Brideshead Revisited* would be a more challenging book if there were any suggestion that Ryder's memories were not altogether to be trusted. In the last analysis, though, the novel is critic-proof; the gracious living and the good meals and fine wines and the splendid architecture and the adultery and the religion and the enveloping aristocratic fantasy provide a potent mixture, which will continue to ensure the book a devoted readership.

In *Sword of Honour* Waugh describes Major Ludovic's very successful novel, *The Death Wish*:

It was a very gorgeous, almost gaudy, tale of romance and high drama ... Had he known it, half a dozen other English writers, averting themselves sickly from privations of war and apprehensions of the social consequences of the peace, were even then severally and secretly, unknown to one another ... composing or preparing to compose books which would turn from the drab alleys of the thirties into the odorous gardens of a recent past transformed and illuminated by disordered memory and imagination.

It sounds remarkably like *Brideshead Revisited*, but as Waugh implies, other works of compensatory fantasy were appearing and finding eager readers. One such was *The Unquiet Grave*, by Waugh's old friend and butt, Cyril Connolly; reading it in January 1945, soon after it was published, Waugh described it in his diary as 'half commonplace book of French maxims, half lament for his life'.[7] Paul Fussell has some interesting reflections in his *Wartime*, treating both *Brideshead Revisited* and *The Unquiet Grave* as instances of a wartime trend towards opulence, lushness, and sensuous richness that crossed literary genres. Colour, in paintings, costumes, interior design, contrasted with the universality of khaki; memories of rich meals and fine wines were an attempt to banish the realities of rationing and dull meagre food; recollections of Paris and Rome and Venice were treasured in the insular prison; even the exuberance of baroque curves like those of the great fountain at Brideshead which Ryder attempted to draw made an assertion against the dominance of the straight line, of men standing to attention, of the ordering of barrack rooms and rows of tents, and, ultimately, of the flight of the bullet.

Fussell sees the most full-throated expression of nostalgia for a richer past, written by a real aristocrat not a worshipper from a distance like Waugh, in the opulent prose of the five volumes of Sir Osbert Sitwell's autobiography which came out between 1941 and 1949, and which were immensely popular. Fussell finds a parallel to Sitwell's richness of language in the poetry of his sister Edith, and in the verbal extravagances of her protégé Dylan Thomas. He suggests that the line continued well after the war: 'the stylistic lushness of Waugh and the Sitwells and Thomas was gratifying enough to deprived audiences to persist as a notable postwar style, from Christopher Fry's *A Phoenix Too Frequent* all the way to the late 1950s, with Lawrence Durrell's *Alexandria Quartet*.'[8]

The popularity of such works as those by Waugh and Osbert Sitwell suggests that they were enjoyed by many readers who had not themselves participated in the forms of life so lovingly recalled; for them these books provided not nostalgia but pure fantasy. The prevalence of aristocratic nostalgia and images of indulgence points to a different facet of wartime consciousness from the egalitarianism of the 'thoughtful corporals', looking for a social democratic future and wider opportunity: the state of mind that Waugh, in a vein of ridiculous snobbery in *Brideshead*, exemplified in the odious young officer, Hooper. We are close to a familiar dichotomy in English culture, that between Cavaliers and Roundheads. In fact, both attitudes co-existed, notably in wartime writing by new and unknown authors, where we find both the tuppence-coloured modes of fantasy and dream narration, and the penny-plain of realistic sketches of service life and straightforward attempts to 'tell it like it is'.

IV

During the war the preferred form for new fiction writers was the short story, or the prose sketch that draws directly on experience but may lightly fictionalize it. If they were lucky, they might get published in *Penguin New Writing* or even in *Horizon*, which published fewer stories and was more choosy, but also came out more often, monthly rather than three or four times a year. But for those who could not make it in these prestigious publications there were many other outlets, in magazines that survived from before the war, or in the irregularly appearing literary miscellanies that were launched subsequently. It is a sobering thought that in the midst of war, despite the effects of bombing and an acute shortage of paper, as well as of other commodities and of labour, there were far more publications where a new writer could hope to place serious short stories than is the case fifty years later.

A good representative selection of material by both new and established writers is to be found in *Short Stories from the Second World War* (1982), edited by Dan Davin. In his introduction, Davin, who himself went through the war in the New Zealand army and wrote stories about it, reflects on the opportunities that war brings the writer—not least, the encounter with a wider range of people and backgrounds than literary intellectuals normally get to know, and the greater sense of the value of human life that is brought by the prospect of an imminent end to it. Davin continues:

Again, if time and privacy can be found, war is favourable in another way to the writer, and particularly to the writer of short stories. The threat of death or maiming or separation is so omnipresent that there is available to him a pervading tension of background. In times of violence the solution of tangled plot by the death of dispensable characters is made more plausible. War abounds in extreme situations swiftly enacted and if the writer can transpose them into art rather than merely transcribe them he need never be at a loss for narrative.

The key words are 'transpose them into art'; in much of the writing produced by extreme situations, including a great deal that found its way into print, the transposition had not been properly achieved, as Davin acknowledges:

In traversing all the contemporary material I could lay hands on I could not help being struck by how often the writers of stories were writing too close to their experience, without the time perhaps to reflect that to record an experience is not necessarily to convey its full meaning to another. Strict adherence to the facts too often had produced the material for a story rather than the story itself. The writers were indeed writing from intense experience—more intense, at least in externals, than peace would normally provide—but they were for the most part young men, untrained in writing but desperately aware of how little time might be left to them; and so in too much of a hurry to be able to give their material enduring form.

The most talented of the writers whose stories arose directly out of military life was, I believe, Alun Lewis. He is better remembered as a poet, but in my view his major gift was for fiction. Lewis was a Welsh schoolmaster, born in 1915, who had done research in medieval history at Manchester University, though his real love was literature. In 1940 he joined the Royal Engineers, served in the ranks for several months, then transferred to the infantry and was commissioned as a second-lieutenant in 1941. He was sent to India and Burma but was never in action; indeed, he was very conscious that his wife, undergoing enemy bombing, was more at risk than he was, and the idea that civilians may be in greater danger than men in the armed forces recurs in Lewis's stories. In one of the strongest, 'They Came', a soldier returns from leave after his wife has been killed in an air raid. (This is the theme of another story in Davin's collection, John Prebble's 'The Soldier Looks for His Family'.) Although he was never in battle, Lewis was a victim of the war, for in March 1944 he was accidentally shot in Burma; the precise circumstances of his death remain mysterious. Lewis shared the left-wing aspirations of many young intellectuals of his generation, in or out of uniform, but with the difference that he was not only a radical, but a Welsh radical, with a profound contempt for the

English ruling class. This is made evident in the title-story of his first collection, *The Last Inspection*, which is savagely mocking about a superannuated brigadier who has emerged from retirement at the start of the war; it is written in the spirit of Siegfried Sassoon's satirical poems in the First World War (Lewis was afraid he might be court-martialled for it).

His finest story, it seems to me, is 'Private Jones', which traces the experiences of a young farm labourer and trapper from deepest rural Wales when he is called up into the army. Among his other stories, 'Night Journey' is a small master-piece. Only half a dozen pages long, it describes one of the long journeys in crowded trains that occur so often in wartime writing, and is confined to the conversation between the occupants of a single compartment: a young officer, who looks exhausted and is travelling without luggage; an air force corporal, his civilian girl-friend, and an aircraftsman he is escorting as a prisoner to a detention barracks; two air force servicewomen, one plain, one sexy; a self-satisfied tank corps sergeant-major; and a cocky private soldier, who boasts of his black-market activities and passes round a bottle of expensive whisky. Lewis had a fine ear for speech, and the ways in which it reveals character; the economy and suggestiveness give 'Night Journey' a Chekhovian quality. Another successful story is 'Ward "O" 3(b)', which is longer and more explicit in its effects. It is set in a ward of a comfortable military hospital in India where four officers of very different types and backgrounds are recovering from wounds or illness and getting on each other's nerves.

Another writer who produced memorable stories of army life was Julian Maclaren-Ross, whose first collection, *The Stuff to Give the Troops*, came out in 1944. He first attracted attention with a short story published in *Horizon*, 'A Bit of a Smash in Madras'. This splendidly comic piece showed sharp descriptive writing, accurate dialogue, nicely paced development, and a keen sense of the absurd. These qualities were to characterize the best of Maclaren-Ross's stories, though Connolly was astonished to discover that he had never been in India and had invented the setting. In his army stories, though, he stayed close to personal experience, usually transforming it into high farce. In 'Y List' and 'I Had to Go Sick', for instance, he presents the military-medical bureaucracy as something out of Kafka rewritten by the Marx Brothers. Maclaren-Ross was a self-absorbed writer who drew fiction from his own life, and whose *Memoirs of the Forties*, though an essential guide to the period, contain an element of fiction. Maclaren-Ross describes meetings with Greene and Connolly and Tambimuttu, editor of *Poetry London*, and provides a memoir of Alun Lewis, whom he had met when they were both in the army; they became friends, but with difficulty, since Maclaren-Ross was a private and Lewis a second-lieutenant, though of avowedly unofficer-like qualities, and Other Ranks and officers were not supposed to mix socially. Maclaren-Ross read Lewis's stories in manuscript, and commented: 'after reading them, the army stories which I myself was trying to write seemed by contrast a joke in rather bad taste. This feeling had worn off by the morning, but I went to bed profoundly dissatisfied with myself and my work.'[9] Maclaren-Ross never properly fulfilled himself as a writer, partly, perhaps, because he would rather talk than write. In the early postwar period he became a familiar figure in London literary bohemia, as dandy, wit, and scrounger; he began to mythologize himself, and the process was continued in accounts by Dan Davin and Walter Allen. Anthony Powell turned him into a fictional character as X. Trapnel in *Books Do Furnish a Room*.

Several writers joined the Auxiliary Fire Service, which later became the National Fire Service. Henry Green was one, and made his major literary response to fire-fighting in *Caught*; he also contributed two stories set in the blitz to *Penguin New Writing*, 'A Rescue' and 'Mr Jonas'. They are strong studies of extreme experience, but more conventional in form and style than the novel. The most striking stories by a fireman that I know of are by William Sansom, who emerged as a writer during the war. 'The Wall' is only three pages long, set in a heavy night of the London blitz; a small group of firemen are playing their hoses on a tall blazing building when the front wall leans outward and starts to fall on them:

Three of the storeys, thirty blazing windows and their huge frame of black brick, a hundred solid tons of hard, deep Victorian wall, pivoted over towards us and hung flatly over the alley. Whether the descending wall actually paused in its fall I can never know. Probably it never did. Probably my eyes digested its action at an early period of momentum, so that I saw it 'off true' but before it had gathered speed.

The wall falls onto the firemen but they are saved by being framed in one of the window spaces. 'The Wall' renders superbly the physical observation of the blazing building and the collapsing wall, together with the process of the narrator's perceptions. Reportage has become art by the precision and intensity of the rendering. Equally remarkable is Sansom's 'Building Alive', describing, in four pages, an incident in the flying-bomb attacks on London in 1944. Elizabeth Bowen called these two stories masterpieces, and wrote, 'A Sansom story is a tour-de-force. It is not written *for* effect, he is dealing *in* it.'[10]

Stories about action, about direct and dangerous encounters with the enemy, or the fires he started, were written for the most part by men. But women were heavily involved in the war too, in the women's branches of the services, or by being directed to work in factories or on farms. They wrote many stories about these situations, or the separation from husbands and lovers, or being bombed, or less violent disruptions of domestic life. A substantial collection of wartime stories by women, *Wave Me Goodbye*, appeared in 1988. One sharply written and original example is Diana Gardner's 'The Land Girl', which first appeared in *Horizon*. It is a sardonic little black comedy, the self-justifying first-person narrative of a spoilt, upper-class girl who is sent to a farm to do agricultural work. She falls foul of the farmer's suspicious, disapproving wife, and works out an ingenious way of being revenged. It achieves more than she bargains for, and the marriage breaks up.

One of the finest of all the stories to come out of the Second World War was in fact written by a woman: Elizabeth Bowen's widely anthologized 'Mysterious Kôr'. She wrote a number of wartime stories that express her patrician Anglo-Irish sense of living in a collapsing civilization, symbolized in the dilapidation, through bombing or neglect, of elegant houses. 'Mysterious Kôr' is artistically purer and more concentrated, so it cannot be easily resolved into recognizable themes; it is poetic and symbolist without being vague or invertebrate. Like many of Bowen's novels and stories, it begins with an elaborate physical description: 'Full moonlight drenched the city and searched it; there was not a niche left to stand in. The effect was remorseless: London looked like the moon's capital—shallow, cratered, extinct. It was late but not yet midnight; now the buses had stopped the polished roads and streets in this region sent for minutes together a ghostly unbroken reflection up.' The spectacular description continues of the city strangely lit by a full moon that makes the blackout useless. Human figures wander into the picture: three French sailors looking for a hostel; two air-raid wardens coming off duty. Then, 'a girl and a soldier who, by their way of walking seemed to have no destination but each other and to be not quite certain even of that'. They look into a park but do not enter it. These are Pepita and Arthur, two of the three main characters of the story. Looking at the moonlit city, she says, 'Mysterious Kôr', and after a moment Arthur picks up her reference, which is to Rider Haggard's romance, *She*. In her conversation, Pepita reveals herself as someone impelled to retreat from the dismal reality of London at war into the fantasy world of 'Kôr'; Arthur is more down to earth. They have had a fruitless evening, unable to get into a cinema, and finding pubs too crowded; the park is too cold and too exposed in the moonlight. They have no choice but to return to the cramped and inconvenient flatlet that Pepita shares with her friend Callie, 'physically shy, a brotherless virgin'. The psychological drama that unfolds when they get there forms the rest of this remarkable story. Bowen was to make a more extended treatment of life in wartime London in a novel published in 1949, *The Heat of the Day*.

V

A discussion of wartime writing needs to take some account of non-fictional prose. George Orwell's pamphlet, *The Lion and the Unicorn: Socialism and the English Genius* (1941), now included in his *Collected Essays, Journalism and Letters*, provides a brilliant series of reflections on English life and civilization; in its most famous phrase, Orwell describes England as 'a family with the wrong members in control'. He is writing as a revolutionary socialist and a patriot. He believes that the war has shown the bankruptcy of capitalism, and that it can be won only if the energies of the people are given a socialist direction: 'It is only by revolution that the native genius of the English people can be set free. Revolution does not mean red flags and street fighting, it means a fundamental shift of power.'[11] Orwell sees the desire for change, whether one calls it revolution or radical reform, as something at large in England since Dunkirk.

Louis MacNeice described *The Lion and the Unicorn* to his American readers as 'the most brilliant piece of pamphleteering I have seen since this war began. He has the defects of the pamphleteer—overstatement, flashy generalization, misdirected truculence—but his analysis of the English character and the social evolution of Britain, and his insistence on *local* factors are a relief after the dogma of the pedants and the jargon of arm-chair reformists.'[12] It is indeed Orwell's emphasis on local factors, on particular forms of Englishness, that now stands out, rather than his ideological analysis. The essay is coloured by a double nostalgia. Orwell recalls the revolutionary, egalitarian society that he had found in Barcelona in 1936, in the early days of the Spanish Civil War, and which remained for him a potent symbol of the possibilities of human liberation. He hopes for something similar in England. But he is also inspired by the idea of England itself, as were so many people at that time. He sets out his sense of what makes England different from other countries in a vivid passage, which is partly analysis and partly nostalgic celebration. He presents isolated fragments of Englishness, like one of Auden's catalogue poems, or the juxtaposed images of a documentary film:

When you come back to England from any foreign country, you have immediately the sensation of breathing a different air. Even in the first few minutes dozens of small things conspire to give you this feeling. The beer is bitterer, the coins are heavier, the grass is greener, the advertisements are more blatant. The crowds in the big towns, with their mild knobby faces, their bad teeth and gentle manners, are different from a European crowd. Then the vastness of England swallows you up, and you lose for a while your feeling that the whole nation has a single identifiable character. Are there really such things as nations? Are we not forty-six million individuals, all different? And the diversity of it, the chaos! The clatter of clogs in the Lancashire mill towns, the to-and-fro of the lorries on the Great North Road, the queues outside the Labour Exchanges, the rattle of pin-tables in the Soho pubs, the old maids biking to Holy Communion through the mists of the autumn morning—all these are not only fragments, but *characteristic* fragments, of the English scene. How can one make a pattern out of this muddle?[13]

Throughout *The Lion and the Unicorn* one finds Orwell's characteristic tension between radical, forward-looking convictions, and a conservative, backward-looking imagination. Like other readers, I find this a creative and fertile tension, but there are those who see it as a sign of confusion and disabling ideological contradiction.

Richard Hillary's memoir, *The Last Enemy*, was widely acclaimed in 1942. Hillary was not a poet, but his book provided a kind of answer to the cry, 'Where are the war poets?', which was a call for a Rupert Brooke rather than a Wilfred Owen. Like Brooke in the earlier war, Hillary was an attractive and talented young man who became a culture-hero. He sees himself at the beginning as one of the so-called Lost Generation at prewar Oxford, young men from well-off families, who had been to public schools, and were at the university to have a pleasant time rather than study hard. They were not as dissolute as the denizens of *Brideshead Revisited*, for they were well aware of the new war that was looming rather than trying to forget one that had not long ended. In

Hillary's account, they—and he—were complacent and self-satisfied, and without strong convictions of any kind; they had nothing to do with the ideals of the Left. Most of them died in the first year of the war, particularly those who became fighter pilots. Hillary served and fought, not in any spirit of patriotic dedication but with what would later be called an existentialist need for self-definition; his stance was neither Left nor Right in political terms, but an anarchistic individualism and resistance to authority.

Hillary writes well about undergraduate life, about rowing in Germany in 1938, about learning to fly, and the experience of aerial combat. He gives a painfully precise account of being shot down in flames, when he was terribly burned on the face and hands, and was temporarily blinded. He takes us through the prolonged and often harrowing process of his recovery, including the extensive plastic surgery he underwent at the hands of the great surgeon, Archibald MacIndoe. Hillary was conscious of the grotesque as well as the appalling aspects of what he went through, and he extracts disconcerting comedy from it. His long recovery was punctuated by the news, at regular intervals, of the deaths in action of his friends. Eventually Hillary was patched up sufficiently well to lead a normal life, though his hands remained claw-like, and he was sent on a mission of a propagandist kind to the United States. Back in England, he insisted on returning to flying, and died in an accidental crash in January 1943.

Hillary could write excellent descriptive prose, and seems to have had almost total recall for the things that had happened to him. He was less sure when he tried to set down his feelings; his failure to convey them adequately means there is a kind of blankness about much of *The Last Enemy*. He was only twenty-three when he died, and for so young a man his book is an impressive achievement. But his writing lacks the sense of implication, of language making surprising but convincing revelations, of feeling being both expressed and ordered, that we find in, for instance, Keith Douglas's journal of the desert war.

Arthur Koestler became a friend of Hillary's, and there is

a character based on him in *Arrival and Departure*. After Hillary's death Koestler published a long memorial essay, which quotes extensively from Hillary's letters. He begins by trying to catch the essence of his friend before he turns into myth: 'Writing about a dead friend is writing against time, a chase after a receding image; catch him, hold him, before he becomes petrified into a myth.' But at the end of his essay, Koestler allows the myth to take over: 'There the man ends and the myth begins. It is the myth of the Lost Generation— sceptic crusaders, knights of effete veneer, sick with the nostalgia of something to fight for, which as yet is not'.[14] The public was ready for the myth, and it was this that made *The Last Enemy* so popular.

Other books gave readers the chance of an escape from the war, to other times and places and more glamorous ways of life, recalled in lush or opulent language; as I have remarked, Sir Osbert Sitwell's autobiographical volumes provided all that, and were gratefully received. But of all the wartime books which exploit the emotions of nostalgia and the desire to escape, Connolly's *The Unquiet Grave* is the most potent and deliberate. It was first published by *Horizon* in December 1944 in a limited edition, under the pseudonym of 'Palinurus', though the author's identity was generally taken for granted from the beginning. In Virgil's *Aeneid* Palinurus was the pilot of Aeneas' ship, who fell overboard, was washed ashore in Italy, and murdered by the local inhabitants. Later Virgil meets his ghost in the nether regions; it had not crossed the Styx because his body remained unburied. The title of the book invokes another literary reference to death and burial. The original 'Unquiet Grave' was an old ballad, of which the *Oxford Companion to English Literature* says, 'a lover laments his dead love for a twelvemonth and a day, at the end of which time she speaks to him from the grave, telling him to content himself and let her sleep.'

The Unquiet Grave consists of journal entries, many literary quotations, mostly in French, and aphorisms and reflections about life and art. In his preface to a later edition Connolly explained the book's origins. There was, first of all, what he called 'a private grief'; that is to say, the separation from his wife, Jean, with whom he had been happy in France before the war, and for which he felt to blame. In the public arena there was the continual struggle against the lies and propaganda of wartime, and the assertion of aesthetic values in the dirt and gloom and dilapidation of London. And above all there was the longing for France: 'To evoke a French beach at that time was to be reminded that beaches did not exist for mines and pill-boxes and barbed wire but for us to bathe from and that, one day, we would enjoy them again.' The book is pervaded by a virulent though beguiling form of 'French 'Flu': 'Peeling off the kilometers to the tune of "Blue Skies", sizzling down the long black liquid reaches of Nationale Sept, the plane trees going sha-sha-sha through the open window, the windscreen yellowing with crushed midges, she with the Michelin beside me, a handkerchief binding her hair . . .'; 'Tout mon mal vient de Paris. There befell the original sin and the original ecstasy—the Cross-Roads and the Island. Quai Bourbon, Rue de Vaugirard, Quai d'Anjou.'

The nostalgia and the self-pity do not become oppressive because Connolly interlaces them with other kinds of discourse. There are outbreaks of ingenious wit: 'The Vegetable Conspiracy: Man is now on his guard against insect parasites; against liver-flukes, termites, Colorado beetles, but has he given thought to the possibility that he has been selected as the target of vegetable attack, marked down by the vine, hop, juniper, the tobacco-plant, tea-leaf and coffee-berry for destruction?' The aphoristic reflections on art and culture can be penetrating: 'Surrealism, like its rival, classical humanism, is too romantic and too anti-industrial for the times. Our world has no use for liberal father or rebellious anarchist son. *Le merveilleux*, with the Sublime of the Humanists, belongs to the nineteenth-century past.' In his aphoristic mode Connolly imitates and invokes the seventeenth-century French moralists, and there are frequent references to the greatest of them, Pascal. In contemplating Pascal, Connolly considers, with seriousness and an element of longing, the claim of religion; he cannot accept it, though he wonders if he might be happier if he could. But he prefers pagan hedonism to Christian asceticism. For him, as for many earlier European thinkers, art remains the supreme source of value and possible happiness. Connolly writes about it with a commitment that goes beyond the usual trivial associations of the word 'aesthete': 'When all the motives that lead artists to create have fallen away, and the satisfactions of vanity and the play-instinct been exhausted, there remains the desire to construct that which has its own order, as a protest against the chaos to which all else appears condemned. While thought exists, words are alive and literature becomes an escape, not from, but into living.'

The Unquiet Grave shows Connolly, in reaction to the collectivist pressures of wartime, turning away from the *de rigeur* socialism of the 1930s (though he was still to give a qualified welcome to the Labour victory in 1945). In loosely political terms, it may be called snobbish and élitist, representing a very different consciousness from the social ideals of Common Wealth and *Picture Post*. Yet it became extraordinarily popular. It was published in a limited edition on handmade paper, and sold out at once. This indicates something about the taste of many readers, and something, too, about the extraordinary persistence of cultural activity; it was the fifth winter of the war, shortages were universal, London had endured the blitz, was recovering from the flying-bomb campaign and was still under attack from V2 rockets. In 1945 *The Unquiet Grave* went into several impressions of a commercial edition from Hamish Hamilton and is still reprinted from time to time. It is a minor and precious work of art, but it is genuinely art and more than just a document of its period. There is a particular sense in which the passing of time has given a fresh resonance to Connolly's nostalgia. With the arrival of mass tourism the civilized pleasures of travel and cosmopolitan living that he longed for in his insular confinement have become even more remote.

An early and very attentive reader of *The Unquiet Grave* was Evelyn Waugh, who went through and annotated the book in Yugoslavia in January 1945. Martin Stannard, in the second volume of his life of Waugh, gives a fascinating account of these annotations, which he describes as 'an odd mixture of brilliant insight, self-delusion and snooty patronage'. Waugh welcomed Connolly's retreat from the Left, but was savagely critical of everything else: his faults of style, his sentimentality, and his alienation from true belief. Waugh's *Brideshead Revisited* was about to appear, and as Stannard suggests, the vehemence of his attacks on *The Unquiet Grave* may have come from his own uneasy sense of how much the two books had in common: 'in many respects complementary works, the same vision from different angles: both nostalgic, both "romantic", both sentimental and deeply prejudiced, yellow with hatred and self-pity.'

Arthur Koestler's collection of essays, *The Yogi and the Commissar*, appeared in April 1945 and provoked immediate attention and some hostility. Koestler had a unique place as a commentator on life and ideas in wartime England; it was natural to him to be called an 'intellectual', whereas for English writers the term was an embarrassment. *The Yogi and the Commissar* includes Koestler's study of Hillary, his reflections on 'French 'Flu', and a number of mainly literary essays. But the more weighty part of the book is the second half, which Koestler wrote specially for it, and which provides an analysis of the nature of Soviet power, then moving into Eastern Europe in the final phase of the war. *Darkness at Noon* had been a classical treatment in fiction of the nature of totalitarian society, and for Koestler the Stalinist system remained unchanged, despite the wartime alliance with Russia and the general enthusiasm for the heroic fight of the Red Army. He insists, despite the attachment to it of many western socialists, that the Soviet Union was in no sense socialist. He provides detailed information about the Soviet mass deportations from the Baltic States in 1940–1, about which nothing was known in Britain; and writing while the war was still on, he gives a remarkably prescient account of the impending division of Europe. Koestler intended to warn readers in the West that though one totalitarian power had been defeated, another had become dominant; there was no great gain for democracy in exchanging Hitler for Stalin, which was to be the fate of millions of Europeans. Koestler was several years ahead of his time; he wanted to warn, but in 1945 he was widely abused for blackguarding an ally, and a 'socialist' ally at that. Nevertheless, his book provided an uncomfortable suggestion of the nature of the coming postwar world.

3

Poets at Home and Abroad

I

Many poems were written during the war, but most of them were quickly forgotten. The outstanding exception is T. S. Eliot's *Four Quartets*, the major poetic work of the early 1940s. Eliot had published the first of the Quartets, *Burnt Norton*, in 1936 as a separate, free-standing poem, without any thought of it being part of a sequence. But early in 1940, as I have described in Chapter 1, he wrote another poem with the same structure, *East Coker*, which instantly became popular. It was followed by *The Dry Salvages* (1941) and *Little Gidding* (1942); the four poems were then published in one volume as *Four Quartets* (1944). For Eliot, as for so many, the war years prompted introspection and retrospection, and the Quartets are his most directly personal poetry; East Coker was the name of the Somerset village from which his ancestor had set out for America in the seventeenth century; the Dry Salvages are a group of rocks off the coast of Massachusetts, which were a feature of his childhood holidays; while Little Gidding was the religious community set up before the English Civil War by the Anglican divine Nicholas Ferrar, embodying the spiritual and cultural traditions to which Eliot had attached himself. The Quartets both look back to Eliot's American origins and confirm his present sense of English identity in a world at war:

> So, while the light fails
> On a winter's afternoon, in a secluded chapel
> History is now and England.

The Quartets are religious poetry of a very high order, but it is a poetry of spiritual exploration and search rather than of dogmatic affirmation. For this reason they attracted reverent agnostics as well as those who shared Eliot's beliefs. They reflected and made coherent a dominant mood; not the political and cultural aspirations for a better world, but the complementary sense of solitude and the need to affirm individual identity and value. In Eliot's sequence the literal darkness of the blacked-out island, and the metaphorical darkness of loss and uncertainty, were echoed and transformed in an individual's journey through his own inner darkness, with a hint of the Christian mystics' dark night of the soul.

Eliot himself thought *Four Quartets* was his major achievement, but those who admired *The Waste Land* and were of a secular frame of mind missed the qualities of the earlier poetry: the parade of dramatic figures, the jazzy vitality, the colloquial speech, the dissection of the urban desolation of the modern world. George Orwell, for instance, was disconcerted by Eliot's new mode, complaining of 'the negative Pétainism which turns its eyes to the past, accepts defeat, writes off earthly happiness as impossible, mumbles about prayer and repentance'. For better or for worse, Eliot—who, unlike more prolific poets, did not repeat himself—had moved on. In his essay 'The Three Voices of Poetry' he had described the first voice as that of the poet speaking to himself or to God; the second is when he is addressing other people; and the third when he speaks through an invented figure, in a verse play or a dramatic monologue. In the Quartets the third voice, so evident in 'The Love Song of J. Alfred Prufrock' or *The Waste Land* or *Murder in the Cathedral*, is absent and the other two dominate. Eliot provides the speech of quiet reflection, of prayer, of song or lyrical invocation, of philosophical argument, together with the tones of preaching and exhortation. I do not find they work equally well; some passages are magnificent, but others are flat and bathetic. The voices may not all be equally convincing, but the locations are real enough, even if they lack the hallucinatory vividness of the settings of *The Waste Land*. We encounter a neglected rose garden, an underground train, the streets of London at the end of a night of bombing, the church of East Coker and the chapel of Little Gidding, the remembered Mississippi and New

England coast. The settings and landscapes have identities and names; at the same time, Eliot gives them a strong symbolic value. The Quartets are, among much else, topographical poetry of a new kind.

In composing the poems Eliot had to face the problem of making them correspond to a template or set form which he had already established for himself. *Burnt Norton* was written in 1935, and made up, as so many of Eliot's earlier poems were, of separate fragments (some of them discarded from *Murder in the Cathedral*). He turned them into a poem with a five-part structure, based on that of *The Waste Land* (which had itself originated in Ezra Pound's editorial manipulation of a mass of fragments). *East Coker* and the succeeding Quartets replicated this structure, in a way that was alien to Eliot's customary way of regarding each new poem as a fresh attempt to deal with intractable experience, demanding a new form of expression. Helen Gardner's informative book, *The Compósition of Four Quartets* (1978), describes Eliot's difficulties with this approach, particularly in writing *Little Gidding*. In general he overcame them triumphantly, but in places a sense of strain persists.

In fact, Eliot's concern with the difficulties of writing is a central preoccupation in *Four Quartets*:

> Words strain,
> Crack and sometimes break, under the burden,
> Under the tension, slip, slide, perish,
> Decay with imprecision, will not stay in place,
> Will not stay still. Shrieking voices
> Scolding, mocking, or merely chattering,
> Always assail them.

Here and in similar passages Eliot is expressing exasperation at the problems of his craft and questioning his poetic vocation as he looks back over 'Twenty years largely wasted, the years of *l'entre deux guerres*'. But they also have an impersonal dimension. As a late poet in the Symbolist tradition, Eliot shares Mallarmé's conviction that 'poetry is made of words', whilst being only too well aware of the vulnerability and insufficiency of words. It is a recurring dilemma in modernist and Symbolist poetics. *Four Quartets* may be, as many readers believe, a triumphantly unified poem; to speak for myself, I prefer to regard it as sequence of twenty sections, not all successful, but still containing some of the finest poetry of the century.

Accounts of English poetry in the 1940s often imply that the political and public concerns of the Auden group disappeared overnight in 1939 when its leader went to America, to be replaced by something variously known as the New Apocalypse or Neo-Romanticism, which adversely dominated the poetry of the war years. There was, indeed, something known as the New Apocalypse, which, unusually for England, was deliberately organized as a movement, with manifestos and several anthologies. But it was never more than marginal in the literary developments of the 1940s. The first anthology, *The New Apocalypse*, subtitled 'an anthology of criticism, poems and stories', was published in 1939. No editors were listed, but the introduction was written by J. F. Hendry. He and Henry Treece were substantial contributors, and the second and third anthologies, *The White Horseman* (1941) and *The Crown and the Sickle* (1944), carried the names of Hendry and Treece as editors. Their initial motive was largely reactive, a desire to dethrone Auden, whose dominance was becoming oppressive by the late 1930s, and to replace him with Dylan Thomas, still only 25 in 1939, but already a cult figure of a provincial genius among young writers looking for new literary directions. In their poetry, Hendry and Treece and other contributors to the anthologies showed themselves obsessive pasticheurs of Thomas's early poetry. Yet there was more involved than simple desire for a change of poetic fashion. In socio-cultural terms there was a reaction against the public-school, Oxbridge and London affiliations of the Auden group, which expressed left-wing sentiments in Establishment accents.

The New Apocalypse rejected political poetry and the idea of collective solutions to public problems. 'Myth' was a key word, and so was 'organic'; it was opposed to the 'machine men', such as Auden and Wyndham Lewis, and upheld individualism, personalism, and anarchism. The admired artists were a mixed bunch, including Leonardo, Shakespeare, Beethoven, Goya, Van Gogh, Picasso, D. H. Lawrence and Dylan Thomas. The New Apocalyptic emphasis on subjectivity, interiority, and irrationality was in tune with wartime sensibility, and so was the taste for mythology, and a mode of religiosity in images and reference. The typical Neo-Apocalyptic poem was likely to be heavily influenced by Dylan Thomas; and beyond him by Gothic writing and the pseudo-medievalism which is prominent in English Romanticism. Ghosts and graves were prominent. Some able poets, such as Norman McCaig, were early on associated with the movement, but most of the poems and stories included in the anthologies were derivative and mannerist and of little intrinsic merit.[1]

The New Apocalypse could not have existed without Dylan Thomas to inspire it. But Thomas kept his distance; he contributed a poem and a short story to the first anthology and nothing to the subsequent ones. During the war his reputation grew and he moved from coterie celebrity to public acclaim. He was writing few poems at that time, but they include some of his most famous, and extravagant claims were made for Thomas's stature as a poet. Thomas was later attacked by Kingsley Amis and other writers of the 1950s, and he now seems subject to a form of benign neglect. He was condemned by hostile critics as 'adolescent', and the posthumous publication of his notebooks gave this charge a literal application; they showed that many of the poems that established his reputation in the 1930s were originally drafted when he was in his teens. The young Thomas was a poet of great gifts and enormous promise: in his sense of rhythm, his capacity to write in complex stanzas and to play sound patterns against syntactical ones, and his feeling for the possibilities and implications of single words. The poems are exciting, and they convey the excitements of adolescence, at the discovery of sexuality and the interaction of the body and nature. They explore process, biological and cosmic, where bodily fluids like blood and semen mingle with tides and clouds and rain. It was a good starting point, but Thomas's problem was how to move on from it.

Thomas may have been the cleverest teenage poet since Rimbaud, and he referred to himself as the 'Rimbaud of Cwmdonkin Drive', though Donald Davie has deplored 'the anachronism of Dylan Thomas playing in 1940 the Rimbaud game for which the right time was 1870'.[2] In his more obscure poems he seems to be attempting the Mallarméan ideal of a poetry that would be as autonomous as music, all suggestion and no statement. It is not easy to know what Thomas had actually read, and he has been mythologized both as a naïve poet of little formal education and one of great learning and intellectual sophistication. It is possible to see his earlier poetry as enacting a separation of word and world, looking back to French Symbolism and forward to structuralism. Some poets have devoted themselves to the ideal of wholly autonomous language—and many poets are perhaps attracted to it at moments—but language is inextricably involved with meanings and reference, which may be neutralized for a time but never completely dissolved. Thomas has always been hauled back to meaning by his commentators, who have made it their business to discover some kind of sense in his poems, even the most riddling.

Thomas's admirers during the war were certainly not concerned with abstruse manifestations of late Symbolist poetics. His basic preoccupations with sex, nature, and religion had a wide appeal, however muffled their mode of expression. Thomas was enjoyed for *sounding* like a poet, for the richness of his language, for the high flights of his rhetoric, the lush density of his imagery, and the splendour of sound. Thomas himself was always endeavouring to recapture the fullness of his earliest inspiration, and ready to strengthen his powers of expression with the steroids of rhetoric. In his later years he escaped from adolescence, once but no longer a potent source of inspiration, by moving back to childhood experience and traditional nature poetry. One can point to 'Fern Hill', 'In Country Sleep', 'Over Sir John's Hill', 'Poem on His Birthday', and 'In the White Giant's Thigh'. These are beautiful poems, but they lack the intensity as well as the obscurity of such earlier work as, say, the 'Altarwise by owl-light' sequence. Wordsworth has replaced Rimbaud. G. S. Fraser, attempting an overall placing of Thomas, has described him as a 'major minor poet', below Thomas Gray, for he has written no long poem of the mature moral interest of the 'Elegy Written in a Country Churchyard', but perhaps on a level with William Collins: 'he might well rank as Collins ranks; he has written some perfect poems, his poetic personality is a completely individual one, he brings in a new note.'[3]

Paradoxically, Thomas's most enduringly popular work was written in prose. This was the radio play *Under Milk Wood*, which was first broadcast early in 1954, soon after his death, and was later adapted for the stage. It is a latter-day pastoral, showing a pageant of colourful characters going about their work and play during the course of one day in the Welsh seaside town of 'Llaregyb' ('bugger-all' backwards). It is influenced in both its detail and its structure by Joyce's *Ulysses*, particularly by the 'Nighttown' section. Nothing very much happens, but the inhabitants have a lot to say about their lives and each other; they are all fluent talkers with a rich range of picturesque or salty speech. Compared with Thomas's early poetry *Under Milk Wood* is undemanding and immediately enjoyable. It is infused with a comic and genial sentimentality, evident in Polly Garter's outcry, 'Oh, isn't life a terrible thing, thank God?'

In 1941, Fraser, as a twenty-five-year old soldier, con- tributed a long introduction to the second Neo-Apocalyptic anthology, *The White Horseman*, which gave the impression that he was much more closely associated with the movement than he was. In fact, he had not met Hendry or Treece when he wrote it, and his point of contact was probably his friend Nicholas Moore, a contributor to the first two anthologies. Neither Fraser's personal tastes nor his poetry were close to those of the Apocalyptics, and his introduction now reads like a smoothly written PR production—Fraser had been a professional journalist before joining the army—rather than a work of committed advocacy. Nevertheless, his brief and somewhat accidental association with the New Apocalypse illustrates his tendency to have a hand in the successive literary movements of his time. In 1941 the fortunes of war sent him to Egypt, where he was involved with the expatriate writers, civilian and military, who got together in Cairo and whose attitudes and practice were closer to Fraser's own mildly Horatian and classical temperament. Then in the 1950s he was engaged in promoting the new poets and novelists of what became known as the Movement. Treece remarked that Fraser 'was considerably more valuable as a critic than as a poet, in which medium he was scarcely Apocalyptic'.[4] This may well be true, but it reflects the limitations of the school rather than Fraser's. His poem, 'S.S. City of Benares', for instance, which appeared in *The White Horseman*, is more accomplished and more affecting than most of the poems in the Apocalyptic anthologies. It responds to the sinking of a ship carrying refugee children; these are the final lines:

> Think what you will, but like the crisping leaf
> In whipped October, crack your thoughts to grief.
> In the drenched valley, whimpering and cold,
> The small ghosts flicker, whisper, unconsoled.

Fraser was indeed better known as a critic than as a poet, but his best poetry is rewarding; his collected poems appeared posthumously in 1981.

Nicholas Moore, like Fraser, was associated only loosely with the New Apocalypse, and neither of them contributed to the third anthology. Moore's poetry was free from the Dylanisms and Hopkinsese and Gothic effects favoured by the true Apocalyptics. His language was simple and transparent, often to the point of seeming *faux-naïf*, and his short lyrics tend to be close to actual song; he was a lover of jazz—see for instance his 'Elegy for Four Jazz Players'—and blues lyrics probably had an influence on him. He was influenced, too, by surrealism, in its whimsical rather than its portentous aspects; and, despite the bans of the New Apocalypse, by Auden. Moore's writing could be facile, and his slighter poems suggest the relaxed charm of the improvising jazz pianist. But he was

also a well-read and competent craftsman, with a classical
education. He was one of the first people in England who
knew about Wallace Stevens, long before any British edition
of Stevens's poems had appeared. Moore contributed a
commentary on Stevens's 'The Woman That Had More Babies
Than That' to *Poetry London* X, and his own poems show
both deliberate imitation of Stevens and a more thoroughly
absorbed influence. 'The Waves of Red Balloons' is dedicated
to Stevens, and begins:

> For all the pretty noise I found him foremost
> The four past masters of the trumpet nothing
> To his green lines of verse, for in them grew
> A subtler form of mastery than theirs.
> His was the elegiac mystery.

Moore's poem, 'The Soldier in the Ice', some hundred lines
long in four sections, shows the effect of the later Stevens. It is
one of the most striking poems to come out of the Second
World War, and as it is so little known I shall quote the whole
of the second section:

> The summer sea stetches blue and placid:
> The summer sky is still, blue with white clouds.
> The trees do not move. Only the small birds move
> And little creatures, the field mouse and the vole,
>
> Or smaller insect on the plantain leaf.
> The soldier stands frozen in ice of wind,
> The soldier stands frozen in ice of thought
> And for a moment here contemplates death,
>
> The death of the mouse, or the rat, or the rabbit.
> Yet this is strange to his habit of mind. He sees
> Fire and fury, familiar blastings and burnings,
> Drills and punishments, mud, and rain, and wet,
>
> A splashed and lifeless face: but this is a dummy,
> A bag of straw that swings with familiar motions,
> A casual scarecrow to keep the jinx away.
> The summer, a cold summer sea stretches
>
> Blue and placid. Cold and lifeless he stands,
> Turning and twisting the cold joints of his hands,

> Fading against the landscape of still trees,
> Watching bird, beast and the summer sky.

During the war Moore contributed to all the little magazines of the day, and published several books and pamphlets of verse; *The Glass Tower* (1944), in which the poems quoted above appeared, was a substantial and impressive collection. Undoubtedly he wrote too much, and within a few years he was largely forgotten, in a peculiarly harsh turn of the wheel of fortune. Moore deserved better. He published occasional poems over the years before his death in 1986, including a remarkable set of free translations of Baudelaire's 'Spleen'; in 1990 a posthumous, rather inadequate selection of his poems appeared under the title of *Longings of the Acrobats*.

The 'New Romanticism' was a broader concept than the New Apocalypse, more a state of mind than a movement. It represented a return to the values and themes of English and European Romanticism—subjectivity, personal vision, myth—and the pursuit of remote or exotic subjects in mythology, history, literature, and art. In wartime culture it aspired to escape from drab or grim reality into a world of the imagination, as opposed to the complementary impulse to render that reality truthfully in writing. A Romantic spirit was dominant in painting, where it achieved more of lasting value and interest than in literature. The nature of the achievement was revealed in the extensive exhibition put on at the Barbican Art Gallery in 1987, under the title, 'A Paradise Lost: the Neo-Romantic Imagination in Britain 1935–55', which included drawings and paintings by Cecil Collins, Henry Moore, Keith Vaughan, Ceri Richards, John Piper, John Craxton, David Jones, Robert Colquhoun, John Minton, Michael Ayrton, Graham Sutherland, Gerald Wilde, and Leslie Hurry.

Close connections between art and poetry were evident in the magazine *Poetry London*, and its associated publishing list, Editions Poetry London. The tenth issue of *Poetry London*, which appeared late in 1944, was a substantial hardback book, widely believed to have been brought out to enable the editor, the colourful M. J. Tambimuttu, to get shot of the many poems he had accepted over the years and not published. (He had great enthusiasm for poetry, but no judgement, so he was likely to publish both the best and the worst he could find.) *Poetry London* X included illustrations by Gerald Wilde to Eliot's 'Rhapsody on a Windy Night' and by Mervyn Peake to Coleridge's 'Ancient Mariner'; among the books published by Editions Poetry London, David Gascoyne's *Poems 1937–42* was illustrated by Graham Sutherland, and Nicholas Moore's *The Glass Tower* by Lucian Freud.

In 1941 the Romantic ideal was invoked in an anthology entitled *Eight Oxford Poets*. The foreword by the co-editor, Sidney Keyes, claimed that the poets included were 'Romantic writers, though by that I mean little more than that our greatest fault is a tendency to floridity; and that we have, on the whole, little sympathy with the Audenian school of writers'. The confident 19-year-old Keyes was misleading in his last remark, though it was in line with Neo-Romantic sentiment, for two of the contributors, Drummond Allison and Keith Douglas, were certainly admirers of Auden. Allison, Douglas, and Keyes were all killed later in the war. Another contributor, John Heath-Stubbs, was medically exempted from military service, and went on to a long and distinguished career as scholar and poet. The subsequent reputation of several of its contributors—another, Michael Meyer, became an authority on Scandinavian drama—has given *Eight Oxford Poets* a greater significance than most collections of undergraduate verse. Wartime Oxford was remarkably rich in poets: Douglas, who had been at Merton, left in 1940; Allison, Keyes, and Heath-Stubbs overlapped at Queen's between 1939 and 1942; Kingsley Amis, Philip Larkin, and Alan Ross were at St John's, the other nursery of literary talent, between 1940 and 1943; John Wain was there between 1943 and 1946, and Amis returned after the war to complete his degree. Philip Larkin has left a sardonic comment on his failure to relate to the Queen's poets, and their high-flying literary Romanticism:

I remember looking through an issue of *The Cherwell*, one day in Blackwell's, and coming across John Heath-Stubbs's 'Leporello': I was profoundly bewildered. I had never heard of Leporello, and what sort of poetry was this—who was he copying? And his friend Sidney Keyes was no more comfortable: he could talk to history as some people talk to porters, and the mention of names like Schiller and Rilke and Gilles de Retz made me wish I were reading something more demanding than English Language and Literature. He had most remarkable brown and piercing eyes: I met him one day in Turl Street when there was snow on the ground, and he was wearing a Russian-style fur hat. He stopped, so I suppose we must have known each other to talk to—that is, if we had anything to say. As far as I remember, we hadn't.[5]

The precocious Keyes and the late developing Larkin were both born in 1922; a reminder of the arbitrariness of the idea of literary generations.

In his *Horizon* editorial for November 1941 Connolly reflected on the state of poetry, conceding a lack of vitality and interest: 'There will always be poetry in England: it is the concentrated essence of the English genius, distilled from our temperate climate and intemperate feelings, and there will always be critics who claim that it is dead. But, poetry is going through a bad patch.' A couple of years later things were looking better. In the March 1944 issue Stephen Spender surveyed volumes of poetry published in 1943, listing, in order of merit, books by David Gascoyne, Kathleen Raine, Lawrence Durrell, Roy Fuller, Peter Yates, C. Day Lewis, Edwin Muir, Geoffrey Grigson, and Terence Tiller. Apart from Yates, these were all poets who stayed the course, continuing after the war to maintain and in some cases to enhance their reputations.

Gascoyne, whom Spender placed at the top of his list, was born in 1916 and was an early developer; he published a novel at the age of 17, and a still useful *Short Survey of Surrealism* at 19. He lived in Paris and became an enthusiastic adherent of the surrealist movement, introducing it in his book to English readers. His own volume of surrealist poetry, *Man's Life is This Meat* came out in 1936. Gascoyne was the only English poet of any talent who made a formal adherence to surrealism, although it was short-lived. At the end of the 1930s he was

still living in Paris, and had become a friend and disciple of
Pierre-Jean Jouve, whose poetry of religious and metaphysical
reflection heavily influenced him. Gascoyne's literary and intellectual formation was un-English; his surrealist apprenticeship
continued to affect his work after he had abandoned the movement; he translated Jouve and Supervielle, and Jouve introduced him to Hölderlin, whom Gascoyne also translated.
Poems 1937–42 contains a sequence of poems written in
French, in memory of the composer Alban Berg. Gascoyne
was affected by existential philosophy, and the painfully
intense consciousness in his poems is somewhat reminiscent of
Sartre's novel *La Nausée*, which came out during Gascoyne's
years in Paris. The first section of *Poems 1937–42*, 'Miserere',
shows him dwelling intently on the passion and death of
Christ, though the approach is that of a poetic mythologizer
rather than an orthodox believer. The opening of 'Pietà' from
this section is characteristic. Intense feeling is conveyed by
clear visual images—a likely inheritance from Gascoyne's
surrealist phase—and emphatic aural patterning contained in
extended syntax:

> Stark in the pasture on the skull-shaped hill,
> In swollen aura of disaster shrunken and
> Unsheltered by the ruin of the sky,
> Intensely concentrated in themselves the banded
> Saints abandoned kneel.

The poems in the final section of the book, 'Time and
Place', convey Gascoyne's anguished response to the events of
1938–40, in a world first threatened by war and then over-
whelmed by it. Despite their intense subjectivity they stay
close to the details of a particular time and place, as in 'A
Wartime Dawn', which I quoted in Chapter 1. Gascoyne
retains his feeling for the physical world even when he is
despairing of it.

The bombing of London and other cities involved poets in
physical danger as well as moral anguish. Eliot memorably
evokes the dawn after a raid in the magnificent Dantean
second section of *Little Gidding*. Stephen Spender was one of

the first poets to write about being bombed, in his 'Thoughts During an Air Raid', written in Spain during the Civil War. He went on to write several good poems about the blitz, such as 'Rejoice in the Abyss', 'A Man-Made World', 'Epilogue to a Human Drama', and 'Air Raid Across the Bay at Plymouth'. Louis MacNeice, having written vivid journalistic accounts, attempted a mythopoeic treatment of the fire raids in 'Brother Fire', 'The Trolls', and 'The Streets of Laredo'. Other poems on the subject are Alun Lewis's 'Raiders' Dawn', Hendry's 'Midnight Air Raid', and Francis Scarfe's 'Lines Written in an Air Raid'. One of the best, Julian Symons's 'Elegy on a City', reflects in generalizing terms indebted to Auden on the experience of seeing London bombed from forty miles away. Symons's friend Roy Fuller wrote 'Soliloquy in an Air Raid' and in the last phase of the war 'During a Bombardment by V-Weapons'; his sardonic and Audenesque 'Epitaph on a Bombing Victim' opens, 'Reader, could his limbs be found | Here would lie a common man . . .'

Just as the bombs physically broke open homes and buildings to reveal their contents, so the shattering of London had the imaginative effect of throwing up famous people and events from the long past of the capital. The fire raids of 1940–1, for instance, reminded poets of the Great Fire of the seventeenth century. Spender expresses the rediscovery of the past in the present in 'Rejoice in the Abyss':

> I saw whole streets aflame with London prophets,
> Saints of Covent Garden, Parliament Hill Fields,
> Hampstead, Hyde Park Corner, Saint John's Wood,
> Who cried in cockney fanatic voices:
> 'In the midst of Life is Death!' They kneeled
> And prayed against the misery manufactured
> In mines and ships and mills, against
> The greed of merchants, vanity of priests,
> They played with children and marvelled at the flowers . . .

Spender is deliberately invoking Blake, one of the greatest of London poets. MacNeice does something similar in 'The Streets of Laredo', which translates the fire raids into the jaunty idiom of a Western ballad:

> At which there arose from a wound in the asphalt,
> His big wig a-smoulder, Sir Christopher Wren
> Saying: 'Let them make hay of the streets of Laredo;
> When your ground-rents expire I will build them again.'
>
> Then twangling their bibles with wrath in their nostrils
> From Bunhill Fields came Bunyan and Blake:
> 'Laredo the golden is fallen, is fallen;
> Your flame shall not quench nor your thirst shall not slake.'

A similar note is struck by Ruthven Todd, who was a Blake scholar as well as a poet, in 'During an Air-Raid':

> On the south side of the city, sheltered from space,
> And from the quivering and menacing sirens,
> I let my mind wander, a kestrel to capture
> The survivors of this holocaust of my Siberian hopes . . .
>
> Alone with this ghost of myself, my body's peak
> In the Pleiades, my limbs longer than Asia,
> And the sun in my loins, I look at this lovely land
> And the places that are precious with people.
>
> Cherishing Lambeth not for its priests and its bishops,
> But for Blake who saw Jerusalem pillared
> In Golder's Green and on Primrose Hill, who, O declare it,
> Discovered the word *golden* in eighteen ten . . .

One of the most famous and frequently anthologized poems to arise from the bombing of London is Dylan Thomas's 'A Refusal to Mourn the Death, by Fire, of a Child in London'. There is, in fact, nothing in the poem actually relating the fire to the blitz, but the connection has always been assumed: Thomas wrote other poems explicitly about the air raids, such as 'Among those Killed in the Dawn Raid was a Man Aged a Hundred', and the more opaque 'Deaths and Entrances' and 'Ceremony After a Fire Raid'. The 'Refusal to Mourn' deserves its fame. Like much of Thomas's later poetry it is assertively rhetorical, but the rhetoric here includes the authentically poetic, as in the final stanza:

> Deep with the first dead lies London's daughter,
> Robed in the long friends,
> The grains beyond age, the dark veins of her mother,

> Secret by the unmourning water
> Of the riding Thames.
> After the first death, there is no other.

William Empson made an admiring close reading of the poem, plausibly suggesting that the poet's 'refusal' is to make war propaganda out of the child's death.[6]

Another celebrated poem about the bombing is Edith Sitwell's 'Still Falls the Rain', with the subtitle, 'The Raids, 1940. Night and Dawn'. She was Dylan Thomas's admirer and patron; both went in for the high style, with rich images and biblical sonorities, and both were described as great poets. In neither case has time endorsed this judgement, and if anything Edith Sitwell had less title to it than Thomas. In her best early poetry, such as *Façade*, she showed a childlike freshness of perception and a sly, pleasant wit. But the poems of the war years aimed at major utterance without any of the qualities needed to sustain it; the result tends to be vaporous and resoundingly empty, despite Sitwell's efforts to draw support from the Bible and other literary sources, like the great lines from the last speech of *Dr Faustus* about Christ's blood streaming in the firmament that she incorporates into 'Still Falls the Rain'. The poem begins:

> Still falls the Rain—
> Dark as the world of man, black as our loss—
> Blind as the nineteen hundred and forty nails
> Upon the Cross.

Julian Symons has reasonably complained of 'the vulgar and too-timely smartness that equates the number of nails upon the cross with the year in which the poem was written'.[7] Spender achieved the effect Sitwell was trying for, more economically and precisely, in the last lines of his 'Air Raid Across the Bay at Plymouth': 'Man hammers nails in Man | High on his crucifix'.

A few years later Sitwell returned to the image of crucifixion in 'Dirge for the New Sunrise' the first of her 'Three Poems of the Atomic Age', which vainly attempt to rise to the vast implications of the dropping of atomic bombs on Japan in 1945:

> Bound to my heart as Ixion to the wheel,
> Nailed to my heart as the Thief upon the Cross,
> I hang between our Christ and the gap where the world was lost.

The idea of crucifixion and the physical image of the wounded Christ on the cross became a common emblem for human suffering in wartime poetry. (As indeed it had been for Wilfred Owen in the earlier war.) David Gascoyne's 'Miserere' poems, particularly the last of them, 'Ecce Homo', make a powerful and sustained use of such imagery. Alun Lewis has a poem called 'The Crucifixion', which attempts to explore the consciousness of Christ as he hung upon the cross. Hendry wrote a sonnet called 'Golgotha', and among the contributors to the New Apocalypse anthologies—perhaps taking their clue from Dylan Thomas's early sonnet 'This was the Crucifixion on the Mountain'—the crucifixion was as common a motif as the frontier had been for the poets of the 1930s.

II

Cyril Connolly remarked that contemporary 'war poets' were not a new kind of creature, simply 'peace poets' who had assimilated the material of war. What they wrote might not be patriotic, but that was a healthy sign, 'for if it were possible to offer any evidence that civilization has progressed in the last twenty years, it would be that which illustrated the decline of the aggressive instinct'.[8] Nevertheless, the journalists and public figures who called for war poets were looking for celebrants of the virtues of patriotism and self-sacrifice, on the model of Rupert Brooke in the previous war. In fact, the idea of the 'war poet' is rooted in the situation of the First World War and is less easily applied to the Second; most of the poetry that emerged from it expressed neither the patriotism of Brooke and Julian Grenfell nor the protests of Owen and Sassoon. The general mood of poets was a stoical acceptance that since the folly of politicians had made the war unavoidable, Nazi Germany had to be defeated, but there should be no conventional heroics, nor any illusions about what war involved or any false expectations about the likely triumphs of the outcome. There was also a feeling on the Left that the war against fascism which began in 1939 was somehow the 'wrong' war; the 'right' one was the Spanish Civil War, which had been lost. This mood could not be identified with the attitudes of either Brooke or Owen, though it was affected by poets' knowledge of the poetry of the earlier war. It is finely caught in Herbert Read's poem, 'To a Conscript of 1940'; Read was a veteran of the Great War, in which he had fought with distinction and been decorated. In the poem he addresses a young soldier in 1940 and recalls the failed hopes of his generation, and speaks of the need to fight without exaltation and even without hope:

> There are heroes who have heard the rally and have seen
> The glitter of a garland round their head.

> Theirs is the hollow victory. They are deceived.
> But you, my brother and my ghost, if you can go
> Knowing that there is no reward, no certain use
> In all your sacrifice, then honour is reprieved.

Few of those regarded now as war poets had the extensive experience of action of Sorley, Rosenberg, Owen, and Sassoon. One who did was Keith Douglas, who served as a tank commander in the North African campaign, was wounded by a mine in Tunisia, and was killed in Normandy in June 1944. Douglas was an admirer of the trench poets of 1914–18 and consciously identified himself with them; in his poem 'Desert Flowers' he wrote, 'Rosenberg I only repeat what you were saying'. He believed that they had already said whatever needed to be said about war, so that 'almost all that a modern poet on active service is inspired to write would be tautological'. Douglas was insistent that a 'war poet' was someone who had been directly involved in fighting, dismissing the effusions of 'clerks and staff officers who have too little to do'.[9]

Nevertheless, some of the best-known poems from the war were by writers who had not experienced action, and deal with the oddities of training, or the boredom of waiting for the next thing to happen. One famous example is 'Naming of Parts' by Henry Reed, who served in the army for only a few months before being transferred to work at the Foreign Office. Another is Alun Lewis's 'All Day It Has Rained . . .':

> All day it has rained, and we on the edge of the moors
> Have sprawled in our bell-tents, moody and dull as boors,
> Groundsheets and blankets spread on a muddy ground
> And from the first grey wakening we have found
> No refuge from the skirmishing fine rain . . .

The poem was widely admired, and in some respects it is an archetypal 'war poem' of the Second World War; it contains no hint of heroics, and the only enemy is 'the skirmishing fine rain' falling on the bored and uncomfortable soldiers. It takes the form of a Romantic meditative monologue, describing the surroundings and then moving into inner consciousness. It is very effective, though I wish it had ended a little earlier than it does, on the lines, 'Tomorrow maybe love; but now it is the rain | Possesses us entirely, the twilight and the rain.' They contain an echo of Auden's 'Spain', but Lewis has integrated it into his response. The remaining few lines make an invocation to Edward Thomas, whom Lewis venerated and whose poem 'Rain' he may have had in mind, but they are somewhat anticlimactic. Lewis's first collection of poems, *Raiders' Dawn*, was well received when it appeared in 1942, and he was quickly deemed to be the war poet that people had been waiting for, a title that was reinforced after his mysterious accidental death in India in 1944; war poets are likely to seem more authentic if they are also victims, as both Brooke and Owen had been. Lewis wrote some attractive poems, but I believe he was essentially a prose writer, whose best short stories show a subtlety and maturity lacking in his verse.

Sidney Keyes was not quite 21 when he was killed in Tunisia in 1943. As I have remarked, he was an early developer and a very bookish poet, who espoused Romanticism at Oxford and bewildered the young Philip Larkin. He was a fluent and sonorous poet who, like the Apocalyptics and Dylan Thomas (though more learnedly and decorously), ministered to the prevalent sense that poetry should sound poetic. He aimed at a fusion of the English Romantic and the Continental Symbolist traditions, and his heroes were Rilke and Yeats. A taste for German poetry was interestingly common during the Second World War, in contrast to the teutonophobia of the First; Alun Lewis addressed a poem to Rilke, and Keith Douglas read him in German while recovering from his wounds. In 'The Foreign Gate', a very ambitious long poem that he composed early in 1942, Keyes quotes in translation the opening lines of the First Duino Elegy—'Were I to cry, who in that proud hierarchy | Of the illustrious would pity me?'—and provides an epigraph to his poem from the Sixth Elegy, to the effect that the hero is strangely akin to the youthfully dead. Keyes was drawn to the exploration of the idea of death that he found in German Romanticism, and in Rilke in particular.

In contrast to Lewis, Keyes had great technical accomplishment, but little contact with the day-to-day world, nor, one suspects, with his own deepest feelings. The accomplishment is in itself somewhat disconcerting in so young a poet, as it did not leave much room for development, and there is an air of the hothouse plant about Keyes's writing. His preoccupation with death can rightly be described as adolescent; but it was also natural enough in an introspective young man of his generation who had no reason to expect a long life. Keyes was killed soon after going into battle and he appears to have written no poems to give him the title of 'war poet' in Douglas's strict and elevated sense. But his death, coming not long after the publication of his first collection, *The Iron Laurel*, made him famous, and his second, posthumous volume, *The Cruel Solstice*, went into several impressions in 1944.

From 1940 to 1943 the only place where British troops were continually in action was North Africa, in the deserts of Egypt and Libya. There the British Eighth Army fought a highly mobile war against the Italians and the German Afrika Korps, with first one side and then the other on top. In mid-1942 the

Germans swept into Egypt and were halted at El Alamein, not far from Alexandria; then in October the British under Montgomery launched a huge counter-attack which thrust the enemy back into Libya. This was the battle in which Keith Douglas took part as a tank commander, having without permission left his safe job behind the lines to rejoin his regiment and take part in the offensive.

Douglas might have become the outstanding poet of his generation. He was also a remarkable prose writer, whose record of his experiences in battle, *Alamein to Zem Zem*, has become a classic war narrative. And as we see from the drawings with which he decorated his writings in prose and verse, he was a gifted artist. Douglas was a complex and contradictory figure. His biographer and editor Desmond Graham sums up his career at his public school, Christ's Hospital, in these words, 'A keen rugby player and swimmer, a dedicated horseman, a resolute opponent of authority and the injustices of boarding school life, a devoted member of the school's O.T.C. [Officers' Training Corps], a generally admired artist, it was as a writer that his reputation was most firmly based.'[10] Graham's biography reveals a strangely divided figure, who was arrogant and snobbish on the one hand, sensitive and considerate on the other. He was unusual among writers in being wholly committed to military life; as a boy of twelve he had written an autobiographical short story which began 'As a child he was a militarist', and he had considered going to Sandhurst rather than Oxford. He joined the army with the conviction, which remained constant, that he would not survive the war; at the same time, he was dedicated to poetry, and intended to become a major poet.

Douglas faced the idea of his own death coolly, and he wrote with detachment about the deaths he had seen on the battlefield, both in *Alamein to Zem Zem* and in his poetry. 'Vergissmeinnicht' is one of his best known poems, describing a dead and decaying German soldier, with the photograph of his girl-friend in the litter around him:

> Three weeks gone and the combatants gone
> returning over the nightmare ground

> we found the place again, and found
> the soldier sprawling in the sun.

The poem offers both a parallel and a contrast to Wilfred Owen's 'Futility', which describes a dead young English soldier. Douglas's coolness, as against Owen's anguish, points not only to a difference in temperament, but to a difference in attitudes to war (even though Douglas regarded himself as a poet in the tradition of Owen). Douglas represses direct emotion, but it emerges obliquely in language and descriptive detail. 'Vergissmeinnicht' has the starkness of a poster, and needs to be complemented by other and rather subtler poems, such as 'How to Kill', where the soldier-poet implicates himself in the fact of homicide:

> Now in my dial of glass appears
> the soldier who is going to die.
> He smiles, and moves about in ways
> his mother knows, habits of his.
> The wires touch his face: I cry
> NOW. Death, like a familiar, hears . . .

One of Douglas's most finely poised poems is 'Sportsmen'—known as 'Aristocrats' in another version—in which he contemplates with equal affection and satire some of the officers he had known, who had once been cavalrymen and preserved their archaic attitudes: 'Unicorns, almost. For they are fading into two legends | in which their stupidity and chivalry are celebrated.' In his best poems, whether those written before he went into battle, or those which resulted from it, Douglas combined intelligence, tough-mindedness and sensitivity. Vernon Scannell has noted in Douglas's poems written in the Middle East, 'their total lack of self-pity and the absolute refusal to indulge in nostalgic evocations of the joys of civilian life, of peace, security, love and home. The best of his poetry is at once passionate and impersonal . . .'.[11] He took the vocations of poet and soldier with equal seriousness; writing poetry was an art, and he despised the sincere but casual scribblers who wrote so much wartime verse. Douglas may have identified with the trench poets of 1914–18 but he learnt from Eliot and Pound and Auden.

Douglas's poetry takes in both the desert battlefield and Cairo, where he spent leaves. The latter is reflected in such poems as 'Egypt' and 'Behaviour of Fish in an Egyptian Tea Garden', while 'Cairo Jag' presents a savage contrast between the two worlds. During the war, Cairo, an ancient, cosmopolitan melting-pot of a city, was host to a large expatriate British population, as well as providing rest and recreation for soldiers on leave from the desert war. Olivia Manning left a vivid account of wartime Cairo in *Fortunes of War*, her autobiographical *roman fleuve*. Manning herself was associated with *Personal Landscape*, a literary magazine run by a group of expatriate poets, Lawrence Durrell, Bernard Spencer, Robin Fedden, and Terence Tiller. These were all civilians in diplomatic or academic posts, but soldiers also contributed, including Douglas, who published several of his best-known poems there. *Personal Landscape* ran from 1942 to 1945, and the last issue contained an obituary of Douglas, written by Bernard Spencer. Among other poets in Cairo, wearing military uniform but working in propaganda or official journalism, were G. S. Fraser and John Waller.

In an article published in 1944 in *Poetry London* X, Fraser attempts to sum up the work of the Cairo poets and establish a contrast between that and the poetry written and published in England. He sees the latter as crisis-ridden, tormented by catastrophe and crying for redemption, while the poetry of Cairo is calm, conscious of the long perspectives of Eastern Mediterranean culture, and the need for restraint and good form in life and art. One pole is romantic, the other classical, and Fraser identifies with the latter, abandoning any allegiance to the New Apocalypse. The school of Cairo, he writes, 'has the poise and something of the sadness of maturity', and quotes from poems by Spencer, Durrell, Tiller, Fedden, and himself. The article is over-simplifying, as such binary divisions always are, but it provides a helpful way of seeing the variety of poetry produced during the war. Certain subjects tend to recur in the work of the Cairo and *Personal Landscape* poets: Douglas and Fraser both wrote poems called 'Egypt'; Spencer wrote 'Egyptian Delta', 'Egyptian Dancer at Shubra',

and 'Cairo Restaurant'; while Tiller wrote 'Egyptian Dancer'
and 'Egyptian Restaurant'. Fraser's 'Monologue for a Cairo
Evening' looks back at that phase of his life and recalls
acquaintances, bringing in Keith Douglas, with a quotation
from 'Vergissmeinnicht':

> And Keith Douglas's shrewd and rustic eyes
> That had endured 'the entry of a demon':
> His poems spat out shrapnel; and he lies
> Where all night long the Narrow Seas are screaming . . .

Hamish Henderson is the other considerable poet of the
desert war, which he survived, becoming in later years an
academic authority on Scottish oral culture. Henderson was an
exotic figure; a Highlander who had been to an English public
school and to Cambridge; a Marxist and a Scottish Nationalist
who was immersed in German poetry and philosophy. I
believe that he approaches Douglas's stature as a war poet,
though he is very much less known. His reputation rests on a
single slim collection, *Elegies for the Dead in Cyrenaica*, which
came out in 1948 and has occasionally been reprinted by small
publishers. In contrast to Douglas's irony and toughness and
cool repression of emotion, Henderson was not afraid to write
rhetorically, to express strong feeling, even to rant, and to
make abrupt shifts between poetic registers. His *Elegies*, in
their range of allusion and their juxtaposition of literary and
demotic speech, are reminiscent, in a small compass, of
Pound's *Cantos* or David Jones's great work from the First
World War, *In Parenthesis*. Henderson has described Hölderlin
as a major influence on the *Elegies*, and they do not disguise their literariness; there are epigraphs from Goethe and
Hölderlin, and from the Gaelic poet Sorley Maclean, who
also served in the desert war. Henderson goes beyond the
recording of simple personal experience, and gives the sense of
a collective voice; he sees the British and German soldiers as
in reality allies against the desert, and ultimately against
death, the great common enemy. The First Elegy begins:

> There are many dead in the brutish desert,
>   who lie uneasy

among the scrub in this landscape of half-wit
stunted ill-will. For the dead land is insatiate
and necrophilious. The sand is blowing about still.
Many who for various reasons, or because
    of mere unanswerable compulsion, came here
and fought among the clutching gravestones,
    shivered and sweated,
cried out, suffered thirst, were stoically silent, cursed
the spittering machine-guns, were homesick for Europe
and fast embedded in quicksand of Africa
    agonized and died.
And sleep now. Sleep here the sleep of the dust.

Henderson's is a decidedly un-English kind of poetry, which
has made some critics uncomfortable. It is more alien than
Douglas's; but they were both poets who had endured much,
and had learnt the lessons of the masters of modernism. They
knew that poetry is not only drawn from experience, however
extreme and unique, but also from other poetry.[12]

Roy Fuller was an excellent poet of a very different but
equally representative kind, a civilian in uniform rather than a
warrior. His first book, *Poems*, came out in December 1939,
so he was able to regard himself—just—as a poet of the
Thirties, and he continued to uphold the diffuse Marxism that
was part of the spirit of that age (though shifting to the Right
in later years). After living through and writing poems about
the London blitz, he was called up into the navy in 1941, and
became a radar mechanic in the Fleet Air Arm. He served first
in England and then in East Africa, without seeing any action.
In the last phase of the war he was back in London with a
commission and a desk job in the Admiralty. Fuller's poems
about his life during those years, collected in *The Middle of a
War* (1942) and *A Lost Season* (1944), are neat, astringent,
and tending to melancholy. They are concerned with loss,
separation, boredom, and their titles tell their own story:
'Waiting to be Drafted', 'ABC of a Naval Trainee', 'Defending the Harbour', 'Saturday Night in a Sailor's Home',
'YMCA Writing Room', and 'Royal Naval Air Station', which
reflects a similar mood to Lewis's 'All Day It Has Rained':

> The piano, hollow and sentimental, plays,
> And outside, falling in a moonlit haze,
> The rain is endless as the empty days.

Then, 'The End of a Leave', 'Good-bye for a Long Time', and 'Troopship'. Fuller subsequently observed that his poems' dominating theme was separation: 'a characteristic . . . of far too much of the verse of the Second World War, particularly in domestic and backward-looking circumstances (though this objection, which even a guilty party like myself held strongly, had to be modified when, after the war, Keith Douglas's collected poems appeared, and later still when Alan Ross collected his revised war poems in *Open Sea*).'[13] In fact, the poems that Fuller wrote in Africa are new in subject-matter if not in tone. In 'The White Conscript and the Black Conscript' he reflects on racial difference; in 'The Giraffes' and 'The Plains', two of his best poems, he looks at African flora and fauna with an Audenesque generalizing eye.

Alan Ross, whom Fuller mentions, joined the navy from Oxford on his twentieth birthday in 1942 and went on to encounter danger and violent action, though he did not treat them in poetry until after the war. Some of his short poems, such as 'Destroyers in the Arctic', are precisely descriptive, but others resemble snapshots that are slightly out of focus. Ross's major work is 'JW51B: A Convoy', a poem of some 500 lines based on his experiences in a naval battle fought in December 1942 off the coast of North Russia in near-darkness and appalling weather. The destroyer he was serving on was shelled and crippled by a larger German warship; Ross had several comrades killed around him and was himself in mortal danger fighting fires. He describes these events superbly in his autobiography, *Blindfold Games*. But in the poem he does not mention his own experience at all, aiming at epical, impersonal treatment of the battle. I believe he succeeds in his aims, and 'JW51B' is an exciting narrative poem. It stays close to significant detail, including the speech and responses of the seamen, whilst conveying the large outlines of the action. Ross began writing it just after the war, but did not complete it to his satisfaction until many years later. It is a unique as well as a powerful contribution to the poetry of the war; at least, I do not know of another treatment of such a subject.

In *Blindfold Games* Ross discusses the problems he found in trying to write war poems, and his response to other poets' work: 'I came to know nearly all Roy Fuller's poems by heart and it would have surprised me then, reading admiringly on my bunk in *Badger* before whisky took its toll or the padre hauled me out for a nightcap, if I had known he was to become one of my closest friends. The poems of separation from his wife and son which he wrote then came to touch me as if I had known them in real life as well as on the page.'[14] This provides an interesting comment on Fuller's own belief that he had written too much about separation. The experiences of Fuller and Ross and what they made of them in poetry were complementary; one was a poet of endurance, the other of action. They were the two best poets of the war to wear naval uniform, though mention should also be made of Charles Causley, whose first book, *Farewell, Aggie Weston* (1951) contained excellent poems and ballads arising from his service in the navy.

4

The Wake of War

I

The war with Germany ended in May 1945; the war with
Japan three months later, much sooner than was anticipated,
following the dropping of atomic bombs on Hiroshima and
Nagasaki. London at the end of the European War was a drab
and dilapidated city, full of gaping bomb-sites and damaged,
patched-up buildings; food was of poor quality, though dis-
tributed in minimally adequate amounts by a complex but fair
system of rationing. Tobacco was very nasty and drink in short
supply. In describing what was supposed to be the London of
1984, Orwell drew heavily on recent memories of the London
of 1944-5, even down to the intermittently descending V2
rockets, called 'steamers' in his novel. People were deeply
weary after years of hardship, danger, and deprivation.
Nevertheless, the infrastructures of civilized life, such as the
postal service and public transport, continued to function well,
and cultural activity, in theatre, ballet, music, exhibitions of
paintings, and the publication of books and literary magazines,
flourished amid all the shortages and physical obstacles.

Despite the prevalent fatigue it was a time of hope; there
was a determination that the wartime ideals of a better and
fairer society would be achieved. After the defeat of Germany
the Labour Party left Churchill's coalition government and a
General Election was called. To its great surprise as well as its
satisfaction Labour won by a landslide majority and Clement
Attlee became Prime Minister. The Labour victory was a
symbolic event of enormous significance: the votes of men and
women in the services were said to have played a crucial part
in achieving it, and it embodied the ideals of the Thoughtful
Corporals and Lieutenant Archers, the readers of *Picture Post*

and *Penguin New Writing* and *Tribune*, the participants in army educational discussion groups and supporters of the Common Wealth Party (which dissolved itself and joined Labour when the war ended). The new Labour Government, it was believed, would continue and make permanent the wartime spirit of classless comradeship and 'fair shares'. In fact, once in office it found itself faced with quite appalling economic difficulties, when the Americans, having captured British overseas markets during the war, abruptly cut off economic aid the moment it ended. The British economy was in ruins—literally so in some cases—after six years of total war, and though an American loan was eventually negotiated the terms were harsh. Labour did what it could to preserve and extend the egalitarian spirit, if at the cost of turning away its erstwhile middle-class supporters, but hopes for a Better Tomorrow were long deferred; rationing and shortages continued for many years.

There are vivid impressions of the quality of life in London in the spring and summer of 1945 in *Europe Without Baedeker* by the American critic Edmund Wilson. He had been sent to Europe in a war-correspondent's uniform to write articles about the conditions and prospects for recovery in Britain, Italy, and Greece. In his accounts of London he describes the physical drabness and weariness of the people, the awful food and squalid surroundings. At the same time he find things to admire, such as the high quality of the London theatre, and he devotes several enthusiastic pages to a performance of Benjamin Britten's new opera *Peter Grimes*, to which Wilson was taken by a highly intelligent society beauty, Mamaine Paget, who later became Arthur Koestler's second wife. Wilson's attempts at fairness did not prevent him being denounced as violently anti-British when *Europe Without Baedeker* was published in 1947. This was probably because of his sharp criticism of the characteristic attitudes of public-school-educated Englishmen (and some Englishwomen) whom he met in professional and official circles in London, and as army officers and diplomats in Europe. He analyses with severe accuracy their effortless assumption of superiority, sometimes accompanied by considerable ignorance, and the tendency to engage in polite needling of other people, especially those considered to be outside the group; for Wilson the politeness of the English professional classes barely conceals an underlying rudeness. The analysis is made with particular point in a section of *Europe Without Baedeker* called 'Through the Abruzzi with Mattie and Harriet', which Wilson describes as a fictionalized version of actual events. Considered as a story, it has latter-day Jamesian overtones; it deals with two young women working in a United Nations rehabilitation team in the ruins of Italy: Mattie, the American is naive but able and intelligent; Harriet, her British superior, though seemingly self-assured, is actually stupid and incompetent.

Wilson clearly suffered from and resented a degree of needling from the British he encountered, but he may have responded too narrowly in ascribing it simply to public-school attitudes. Arguably, the whole of English society is permeated by an adversarial spirit—evident in law and politics and sport—which sees life as a matter of being continually on top or 'one up'. It was memorably described some years later in the books of Stephen Potter, *Lifemanship* and *Oneupmanship*. Still, Wilson's attack on the styles and assumptions of the English ruling class was sufficiently sharp to annoy reviewers and opinion-formers, and attract the charge of being anti-British. In fact Wilson's hostile analysis, made by an observant outsider, anticipated by a few years the attacks made from within English culture by the new writers of the 1950s, who came from grammar schools and were in revolt against the public-school ethos.

Peace and war are antithetical states. Yet the immediate postwar years, which coincided with the period in office of the Labour Government from 1945 to 1951, in many ways continued the wartime atmosphere. The enemy may have been defeated, but the weary public was enjoined to Work or Want, and to sacrifice domestic needs for the export trade. Physical reconstruction was slow to get under way, rationing was in some respects more severe than in wartime—bread was even rationed for a time, which it never had been in the war—and egalitarianism was imposed from the top, by fiscal policy. Foreign travel was severely limited by exchange controls, much to the resentment of writers and artists who looked forward to resuming the remembered *douceur* of Continental life. Nevertheless, once the war ended the literary situation changed. The lonely exiles in uniform did not read so much after they came home and were caught up in work, family, hobbies, sport, and all the other possible forms of normal sociability. Aspiring amateur writers were deflected from literary ambitions when they returned to civilian pursuits. *Penguin New Writing* had to drop a regular feature of reportage called 'The Living Moment' when the war was over and contributions to it declined.

Editors and critics in the late Forties tended to see the contemporary literary scene as exceptionally barren, and it is usually presented as a dull and empty period. It is certainly true that new writing was very slow to appear; some promising talents had died in action, and others fell silent in peacetime. Nevertheless, writers who were already established—especially novelists—produced outstanding work, and the years between 1945 and 1951 were much more interesting and productive than is often thought. Even the Year Zero of 1945 saw the publication of two major works: Henry Green's *Loving* and Orwell's *Animal Farm*.

Like other novelists at the time, Green focused closely on personal relations, against the background of war. In *Caught* he had traced the lives of a handful of people in London during the first year of war, culminating in the blitz. *Loving* takes place at much the same time—the bombing is happening offstage—but in a world remote from the war; the setting is a country house in neutral Ireland and the characters are the English servants of the Anglo-Irish family who own the house. It is Elizabeth Bowen territory, seen from below stairs. The servants form a remarkably isolated group; cut off from wartime England, suspicious of the Catholic Irish country people around them and worried by rumours of an IRA attack and a German invasion. The dominant figure is the cunning and intriguing Charley Raunce, the head footman, who gets himself made butler after the death of the old butler at the opening of the novel. As one expects from Green, *Loving* is disconcertingly original; in its situation, its language, and its attitudes and off-hand treatment of the war. Green seems fascinated by the oddities of his characters, and they fascinate the reader; Charley, for instance, spies on and systematically cheats his employers but punctiliously sends money to his mother in England. He spreads anxiety about the possible invasion: ' "And what about the panzer grenadiers?" he asked. "When they come through this tight little island like a dose of Epsom salts will they bother with these hovels, with two pennorth of cotton? Not on your life. They'll make tracks straight for great mansions like we're in my girl." ' *Loving* has realistic elements, but the realism is interwoven with a fairy-tale element that becomes explicit in the final words: 'they were married and lived happily ever after.' In fact it resists categorization; it is wonderfully comic, but also delicately lyrical, as in the scene in which Raunce comes across the housemaids Kate and Edith dancing together to a gramophone under the great chandeliers in a deserted ballroom: 'two girls, minute in purple, dancing multiplied to eternity in these trembling pears of glass.'

In 1947 Patrick Hamilton published *The Slaves of Solitude*, which though less subtle and various than *Loving* is one of the best novels of its period. It looks back to 1943, to a genteel boarding house in a Thames-side town based on Henley. The season is midwinter; there seem to be few hours of daylight, and most of the action takes place behind the drawn blackout curtains of the boarding house, in the dark streets or in pubs. In *Hangover Square* Hamilton had shown his interest in lonely and alienated middle-class figures and he continues it in this novel; the heroine of *The Slaves of Solitude* is the likeable but shy Miss Roach, a 39-year-old secretary in a London publishing house; she has few friends but is not so tragically isolated as George Harvey Bone. She is courted in a half-hearted way by an amiable but inconsequential American officer of about her own age, Lieutenant Pike, who claims to

own a laundry in Wilkes Barre, Pennsylvania. He proposes to
her and she is tempted to accept though she does not love him;
but then she discovers that he proposes to practically every
woman he meets. Hamilton shows the exciting but unsettling
effect the Americans, who had plenty of money and access to
seemingly unlimited drink, had on the sedate life of the town,
particularly its women. (American soldiers, British men
commonly complained, were 'over-paid, over-sexed, and over
here'.)

But the main drama goes on in the boarding house, where
Miss Roach is bullied and tormented by one of the other
inmates, the elderly retired businessman, Mr Thwaites, who is
later aided and abetted by Miss Roach's treacherous friend,
Vicky Kugelmann. Mr Thwaites is one of the most splendidly
nasty characters in modern fiction, and Vicky is not far behind; Hamilton has them satisfactorily seen off at the end. The
novel's semi-happy ending means that one can regard it as a
comedy, despite Miss Roach's earlier humiliations. Hamilton
makes a remarkably sensitive exploration of the lonely lives
of the respectable middle-aged, revealing comic as well as
poignant aspects. The novel's principal limitation is an excess
of authorial intervention, though sometimes this works well, in
ways that recall the sharp insights of Jane Austen or George
Eliot: ' "But, my dear, this is *marvellous*!" said Miss Roach,
that slight film coming over her eyes which comes over the
eyes of those who, while proclaiming intense pleasure, are
actually thinking fast.'

At the end of the book Hamilton looks ahead to the 'little
blitz' of February 1944, when for a few nights the Luftwaffe
resumed the bombing of London; and this episode provides a
climax in Elizabeth Bowen's *The Heat of the Day* (1949), a
novel which finely evokes the atmosphere of wartime London:

They had met one another, at first not very often, throughout that
heady autumn of the first London air raids. Never had any season
been more felt; one bought the poetic sense of it with the sense of
death. Out of mists of morning charred by the smoke from ruins each
day rose to a height of unmisty glitter; between the last of sunset and
first note of the siren the darkening glassy tenseness of evening was

drawn fine. From the moment of waking you tasted the sweet autumn not less because of an acridity on the tongue and nostrils; and as the singed dust settled and smoke diluted you felt more and more called upon to observe the daytime as a pure and curious holiday from fear. All through London, the ropings-off of dangerous tracts of street made islands of exalted if stricken silence, and people crowded against the ropes to admire the sunny emptiness on the other side.

This is richly evocative, and the final image is both poetic and precise. But the poise of Bowen's prose is precarious, her flights sometimes collapse into purple passages, and taken as a whole the novel seems to me a very interesting failure. The central character, Stella Rodney, is an elegant widow in her early forties, comfortably off but not rich, engaged in confidential war work in a government office, and renting a furnished flat in central London, whose decoration and fittings, it is made clear, do *not* represent her own taste. (Bowen is supremely the novelist of interior decorating, considered as an index of sensibility.) Stella has a son in the army, and a lover, Robert Kelway, a few years younger than herself, who was invalided out of the army after being wounded at Dunkirk and is now working in the War Office.

The central action of the story, both physical and psychological, arises from the discovery that Robert is passing information to the enemy. Stella is told this news by a security man known as Harrison. She is warned to stop seeing Robert, who is under surveillance, but not to say anything to him. Stella refuses to believe it, but Harrison insists; then he suggests that if Stella sleeps with him Robert will be left alone. None of this strikes me as credible; or to put it differently, the two men concerned are not sufficiently present or alive for one to have any belief in them or what they are supposed to be. Harrison is an evasive, awkward figure, who seems to have strayed out of a Graham Greene novel, though Greene would have made him a little seedier. Robert is a complete cipher; eventually he admits his guilt to Stella and comes out with an inchoate attempt at a fascist justification, but none of it carries conviction. The Marxist critic Alan Sinfield believes *The Heat of the Day* is 'even-handed' about fascism, but I cannot find

anything like that degree of ideological coherence in it. Bowen
appears out of her depth, in dealing with ideas and in the basic
plausibility of the story. There were certainly a few fascist
sympathizers around in wartime England, but no Nazi agents
were ever found in official employment, though as we know
there were a number of Soviet ones.

Despite the central failure, there are some rewarding sections in *The Heat of the Day*; the imaginative energy goes into
presenting places and topics that are peripheral to the main
story, like Stella's visit to Robert's home in the country, where
she meets his formidable mother and elder sister, and where
Bowen engages in some nice social satire and a characteristically sharp analysis of the domestic decor; and particularly
the author's brief return to the physical and emotional territory where she felt most at home, when Stella's son Roderick
inherits an Anglo-Irish country house and Stella travels to
Ireland to inspect the property and meet the servants. This
is one of the best episodes, where Bowen approaches the
material of Green's *Loving* but does not overlap with it;
however similar their subject-matter these novelists inhabit
different worlds. Stella is a convincing character, unlike the
men; her presentation by Bowen as a woman of the finest taste
and sensibility betrayed by a gentlemanly fascist reflects the
author's elegiac feelings for a shattered world of true values
and behaviour.

Graham Greene returned to novel-writing with *The Heart
of the Matter* in 1948 and *The End of the Affair* in 1951. In
terms of Greene's own development they represent a continuation of the Catholic preoccupations which had entered
his fiction in *Brighton Rock* and were continued in *The Power
and the Glory*; they helped to bestow on Greene the title of
'Catholic novelist', though it was one that he always resisted.
The Catholic issues were more precise and demanding than in
The Power and the Glory, which had the wide human appeal
of the story of a hunted man struggling to fulfill his mission
under political persecution. In the two later 'Catholic' novels
questions of marriage and adultery become central, for the
first time in Greene's work. In *The Heart of the Matter* the

Catholic police officer, Major Scobie, serving in a disagreeable West African colony during the war, is a good but flawed man, like a traditional tragic hero though without the stature. Scobie's fatal flaw proves to be pity, which leads him into deeper and deeper trouble; first to adultery and then to compliance in unprofessional practice and, perhaps, in murder; then to receiving the sacrament in a state of mortal sin, and on to the ultimate and technically unforgiveable sin of suicide. Whether Scobie is damned or not is a question which has attracted much discussion among Catholic readers. But there are other arguments about him: is he a self-deceiving neurotic, or a decent man for whom everything goes wrong and who becomes tragically trapped in a course of action from which, in ordinary human terms, there is no escape? Scobie is a humble man, almost without any sense of self-worth, but he can also seem remarkably arrogant, as when, at the end of his sad life in which he has been moved by pity for his wife and mistress, he goes on to pity God. In the end Scobie, with all his torturing dilemmas, seems too good to be true, or too complex to be convincing; *The Heart of the Matter* is something of an over-egged cake. What most stays in the mind is the strong evocation of West Africa in the Second World War, where Greene had been an intelligence officer: the intensities of heat, the rats, the cockroaches, and a prevailing sense of physical and moral corruption. The war is a background presence that intrudes at intervals, as when the girl who is to become Scobie's lover is brought ashore from a torpedoed ship.

In *The End of the Affair* the setting returns to London. The time-scale is fluid; there is a brief glance back to the last prewar summer of 1939, but most of the action occurs during the flying-bomb raids of 1944, and in the first few months of peace. The presentation of London is more direct and prosaic than in Greene's novels of the 1930s, which were full of elaborate metaphors and cinematic effects. In fact, *The End of the Affair* was to be the last novel by Greene with an English setting for many years. It is also the last of his so-called Catholic novels, and in some respects it is the most overtly religious; God becomes a character in the story. Nevertheless, it is a new kind of novel for Greene. It is the first-person narrative, told by Maurice Bendrix, a novelist and an unbeliever, of how he lost his lover Sarah Miles, the wife of a civil servant, to God, whom he ultimately has to acknowledge as a successful rival and a more powerful maker of plots. In unfolding Bendrix's story Greene is indebted to Ford Madox Ford's *The Good Soldier*, a novel he greatly admired. *The End of the Affair* is a transitional work in Greene's development. It is very Catholic in its implications; at the end of the story it is suggested that the adulterous Sarah may have become a miracle-working saint after her death, a suggestion likely to disturb agnostic readers just as the emphasis on sex upset devout Catholic ones. At the same time the character of Bendrix looks forward to Greene's later work. He is an isolated figure, like Pinkie and the whisky priest and Scobie, but he does not have their spiritual dimension. Rather, he is an early version of the lonely, cynical, sometimes embittered observers of the human spectacle who provide the central consciousness of Greene's novels from the mid-1950s onwards: Fowler in *The Quiet American*, Querry in *A Burnt-Out Case*, Brown in *The Comedians*. After the religious phase of his work Greene became concerned with politics and ideology in the Third World, and his personal belief changed from a tormenting orthodoxy to what he later defined as the position of a 'Catholic agnostic'.

The novels I have so far discussed in this chapter are by writers whose names are familiar and whose work tends to be kept in print; they all also look back to the Second World War, either as a background or a focus of action. William Sansom's *The Body* is an exception on both counts. He emerged as a writer during the war and quickly became famous, especially for his short stories about the blitz. By the late Forties Sansom was regarded as one of the leading writers of his generation; his first novel, *The Body*, was widely acclaimed when it was published in 1949. Since then he has become quite forgotten, in an extreme instance of the fickleness of literary taste. A fleeting reference to the atomic bomb places *The Body* notionally in the postwar period, but the atmosphere and setting, in comfortable bourgeois surroundings in a spacious, leafy North London suburb, recall the more prosperous aspects of life in the 1930s, with a few hints of the Edwardian era.

The Body contains very little action. It is a study of the obsessive jealousy of the narrator, Henry Bishop, a semi-retired businessman who is devoted to his garden and likes reading but is generally rather idle. He is only 45 but very set in his ways, as his wife Madge complains. He presents himself as an Edwardian survival, a combination of Prufrock and Mr Pooter: 'I must be equally fair and confess myself probably rather a dull man. It annoys her, I know, that at week-ends I wear old-fashioned tweed knicker-bockers and stockings. I am sure she would rather I was more athletic, and that I had not a jumpy way of walking on my toes.' But Henry's narrative is anything but dull. Sansom's great quality as a writer was an ability to convey experience with hallucinatory vividness, whether it was firefighting in the blitz, or the minutiae of suburban life. The novel opens with a striking instance, as we see Henry in his garden determinedly preparing to drench a fly with insecticide: 'To hold the syringe gently, firmly but delicately—not to squirt, but to prod the sleeper into wakefulness with the nozzle, taking care to start no abrupt flight of fear. Only to stir a movement, to initiate a presence from such a dead sleep. Gently, gently—lean thus into the ivy, face close into the leaves, bowed in yet hardly daring to breathe . . .' Soon afterwards Henry becomes aware that his new next-door neighbour is attempting to peer at Madge in the Bishops' bathroom. The seed is sown of the jealousy that consumes Henry throughout the novel. The neighbour, Charles Diver, is a bounderish car-salesman and a ripe saloon-bar bore, though drawn by Sansom with curious affection. Henry is wrong in his suspicions, and there is a good deal of comedy in perceiving how wrong he is; at the same time the power of his obsession is so intense as to be, despite everything, convincing. There are some passages in which the combination of loathing and fascination is reminiscent of Sartre's *Nausea*:

In the fresh morning air, in the still room without fire or light, in that motionless new grey daylight I sat and stared at the blacklead. After a few minutes, long minutes, I remember my eyes moving nearer to my boots. Nothing stirred—but in the stoneset solitude I suddenly grew conscious of my living body. Inside those black boots there were feet and toes and on the toes greyish-yellow hairs. There was a corn on one toe, a patch of hard skin along the side of the other foot. Inside the boot, inside the sock, there was life. And in this knowledge I understood clearly how all the time, motionless in a motionless room, my body was slowly, slowly falling to pieces. A gradual, infinitesimal disintegration was taking place. Nothing could stop it. Pores that once had been young were now drying up, hairs were loosening in their follicles, there was an acid crusting the backs of my teeth and my stomach.

As we read, we are caught up in Henry's consciousness but not to the point of absolute claustrophobia; there is relief in moving out of it at intervals to Sansom's superb renderings of the physical world: of suburban gardens and interiors, pubs and streets, and—in a nice Edwardian touch—a boating trip up the Thames. *The Body* is a mannered book, which deliberately works in a narrow compass; but what it achieves within it is remarkable.

Though established writers were producing good work in the second half of the 1940s, new ones were slow to appear. Then in 1949 Angus Wilson published his first book, a collection of stories called *The Wrong Set*, which was enthusiastically received. It quickly ran into several impressions, and was followed by a second collection, *Such Darling Dodos* (1950). Wilson was welcomed as the new writer that everyone had been waiting for. He was already 36 when *The Wrong Set* came out, and the dominant consciousness of his writing seemed even older and cynically wiser than that. Wilson's subjects were from the more raffish and insecure ranks of the upper middle class: genteelly alcoholic widows in small private hotels, ex-officers full of despair and violent opinions, ambitious, anxious scholars, despotic liberal intellectuals, fading male or female seducers, violently possessive mothers, and do-gooders who do more harm than good. Wilson was fascinated by the skeletons in the cupboards of supposedly respectable families, or those that self-indulgently proclaim themselves to the world as 'crazy'. He had antecedents in E. M. Forster and Christopher Isherwood, and beyond them in a long line of English writers of sharp observation and realistic social comedy; but he was also interested in violence, occasionally physical—as in that horrifying story, 'Raspberry Jam'—but more often moral and psychological.

Wilson has some affinities with the novelists I have been discussing, particularly Patrick Hamilton, another keen analyst of middle-class decay. If Hamilton and Henry Green fit into the large category of social comedy, one might apply 'social melodrama' to Elizabeth Bowen and Graham Greene. William Sansom has elements of both, and so does Angus Wilson. He shares the retrospective stance of those writers, which was to be the dominant one in English fiction for several more years. In some stories he goes back to the early Thirties and even the Twenties; in one of the best of them, 'Union Reunion', Wilson—who spent his childhood in South Africa—describes a lavish but ultimately grim family party of prosperous South Africans in 1924. Other stories are set in the wartime bureaucracy, where Wilson no doubt draws on his service in the Foreign Office. But he provoked a particular shock of recognition by his feeling for the drab and divided social and mental climate of the immediate postwar years, when the exhaustion of war was giving way to resentment at the slowness of recovery.

Wilson has a keen eye for significant details of appearance and an equally acute ear for the revealing qualities of speech, in vocabulary, idiom, and pronunciation. But he also has a sense of the typical, the capacity to see a situation or an individual as representative of a historical moment. For instance in 'A Story of Historical Interest', Lois Gorringe, an unmarried middle-aged woman, has been wearing herself out in nursing her sick elderly father; he is a charming but un-scrupulous Edwardian roué, whose illness, it appears, may be the result of syphilis picked up while he was sowing the wild oats of his youth. The story takes place at the time of the

Munich Crisis of 1938, when for the first time England was seriously threatened by war. Finally, when her father has had to go into a nursing home and is reported close to death, Lois decides that his condition is 'only of historical interest', picking up a phrase the doctor had used in his diagnosis. It is a fine story, very responsive to the political climate of the time.

Wilson usually writes well about women, especially those who are no longer very young; they may be victims of their surroundings or of tyrants trying to dominate them, but in either case they are drawn with sympathy and insight. The title story of *The Wrong Set* is about Vi, who plays the piano in a seedy nightclub and takes an interest in her nephew, an undergraduate at London University. She is horrified to discover that he has fallen among left-wingers, the 'wrong set' of the title. Vi is contrasted with her nephew's landlady, a staunch traditional Labour supporter, who is wryly intrigued to discover that Vi is a Tory. At the same time she is cross that her son and daughter have gone to a Communist demonstration: '"To make trouble for the Government they put into power" said Mrs Thursby drily. "It makes me very angry sometimes. It's taken us forty years to get a real Labour Government and then just because they don't move fast enough for these young people, it's criticism, criticism all the time".' This brief story, no more than ten pages long, nicely catches the currents of feeling of the late Forties. On the one hand there were Conservatives and the middle classes, furious at confiscatory taxation, continued rationing, and exchange control regulations which restricted foreign travel; all the things which meant that they were never going to recover their prewar way of life. On the other hand, those on the Left were angry when the practical difficulties and compromises faced by a Labour government in economically crushing circumstances meant that the brave hopes of the war years were disappointed.

Wilson cannot be called a political writer, but he is sensitive to the personal and social divisions that accompany or underlie ideological clashes, and the comedy they can provoke. He had a good insight into the way that various groups felt threatened or overtaken by new historical forces. Under a Labour

Government *rentiers* and all those on the Right were fearful of being ground down by socialism. But before long there was a right-wing ideological counter-attack, which disturbed those who regarded themselves as progressives. This was the theme of the title story of *Those Darling Dodos*, where a couple of middle-aged Oxford Fabians encounter a young ex-officer student and his wife, and are disconcerted by their Conservative convictions and strong support of organized religion. In one story, 'Realpolitik', Wilson provides a remarkable anticipation of events in the Thatcher era forty years later: the brash new-broom director of a provincial art gallery makes it clear to the assembled staff that promotional rather than scholarly or artistic values are to prevail henceforth, and privately plans to get rid of most of them. Wilson's own convictions remain those of a liberal and a progressive, but he shows that no one ideological position has a monopoly of folly and moral blindness. His capacity for exposing these things was developed at length in his novels, *Hemlock and After* (1952) and *Anglo-Saxon Attitudes* (1956). In the latter, Wilson undertakes a Dickensian analysis of the condition of England over a period of forty years, with a wealth of grotesque and comic characters. It remains one of the best English novels of the 1950s, though in his sustained fiction Wilson remains a short-story writer at heart, stronger on anecdotal detail than on overall design.

II

Some writers looked for larger topics than the psychological and moral cross-currents of middle-class life, and were conscious of more demanding historical themes than the fissures and uncertainties in English society. There were, above all, the enormous implications of the splitting of the atom and the use of atomic bombs on Japan in 1945; first there came a sense of awe and horror at what had happened at Hiroshima, and then, with the development of the hydrogen bomb in the 1950s and worsening international relations, a pervasive fear of the likelihood of nuclear war and global extinction. A separate though related matter was the Soviet threat. The West and the USSR had been allies against Germany, but after Nazi totalitarianism was defeated the Soviet version, under the paranoid despot Stalin, continued and embarked on what was widely regarded at the time as a course of world domination. The British also had to come to terms with the dissolution of the Empire and the resultant uncertainty about Britain's place in the world. India became free in 1947, and independence was granted to most of the colonies during the 1950s and 1960s, but delusions about Britain's imperial role persisted for a long time.

Fiction that directly confronts major historical events and issues can, if it is good enough, preserve imaginative power and what James called 'felt life'; Tolstoy did so in *War and Peace*, and kept his general ideas about historical determinism for the final chapter. But for lesser writers such subjects may lead either to the discussion novel or to reportage; politics is, in fact, often best treated in the modes of fable and allegory. Arthur Koestler had shown himself to be a political novelist of genius in *Darkness at Noon*, and he continued to invoke and interrogate history in his subsequent novels. But their interest is more historical than literary; the comparison with Conrad that *Darkness at Noon* had evoked was no longer appropriate. His *Thieves in the Night* (1946) is about Zionist settlers in Palestine in the late 1930s, against the background of Nazi persecution in Europe and Arab resistance to Jewish settlement. The characters tend to be cardboard cutouts: intelligent, heroic Jews; stupid, dishonest, and murderous Arabs; polite, devious British administrators. Koestler gives the conflict between Jews and Arabs a sociological dimension, as a clash between the forces of modernization and those of a rigid traditional culture. He brings in a lot of the actual history of that time and place, and shows the intractability of the conflicts in what later became Israel, which have not changed their essential character in more than half a century.

In *The Age of Longing* (1951) Koestler writes of France in the near future, as the Cold War is on the point of turning hot. A Soviet invasion of the West is feared and expected, and in the last pages of the novel it happens. The novel is fast-moving journalistic fiction, recording events but without much imaginative life. *The Age of Longing* is a deeply pessimistic book, pervaded by the conviction that the West has lost the will to defend itself against the threat of Soviet totalitarianism. Its satire of French intellectual life reflects something of Koestler's own situation: *Darkness at Noon* had made him deeply unpopular with the pro-Communist French Left, and there is a hostile portrait of him as 'Victor Scriassine' in Simone de Beauvoir's proprietorial novel about Parisian intellectuals, *The Mandarins*.

The Age of Longing is a 'futurological' novel, looking at the state of the world at a later time, whether near or remote; there were to be several instances in the late Forties and early Fifties. This form may overlap the genre of science fiction, though it is not primarily concerned with scientific change and technological innovation; or it may be written as a fable. Anxiety about nuclear war, in particular, provoked such writing, whether as science fiction or in work by mainstream writers. Aldous Huxley's *Ape and Essence* (1949) is an extreme, melodramatic instance, presenting in the form of a film-script a world after a nuclear war which has reverted to brutal but ritual barbarism; babies born with genetic malformations caused by radioactivity are ceremonially sacrificed.

Orwell's *Nineteen Eighty-Four* (1949) is the most famous futurological novel of this period; its subject is not nuclear war or its aftermath, but totalitarianism. It is perhaps best approached via *Animal Farm* (1945). After producing a series of painstakingly realistic novels in the 1930s, Orwell accepted that in fiction his real talent lay in fable and the projection of imaginary worlds. He wrote *Animal Farm* towards the end of the war and had difficulty in finding a publisher, since its anti-Soviet theme was seen as hostile to the wartime alliance. Because of these delays it did not appear until the summer of 1945, after the end of the European War, which was probably fortunate timing, as pro-Soviet sentiment was cooling by then. It employs the ancient form of the beast-fable to provide a satirical history of the Soviet Union, from the Revolution of 1917 to the Quebec Conference of 1943. Since then *Animal*

Farm has become a literary classic, and appeals to readers who know nothing of its historical and political point of departure. It is, to put the matter at its simplest, a piece of superb story-telling, where Orwell shows the powers of a major writer by knowing what to leave out as much as what to put in. The animals are convincing as animals whilst representing for historical figures and forces, such as the pigs Napoleon and Snowball, who stand for, respectively, Stalin and Trotsky. Not for nothing has *Animal Farm* become a popular children's book while remaining a serious political satire. This has long been true of *Gulliver's Travels* by Orwell's hero Swift, and *Animal Farm* stands the comparison well; both works show how literature may originate in a particular historical situation but subsequently reveal fresh powers and possibilities in its reception by later readers.

Considered as a satire, *Animal Farm* contains an element of ambiguity, like much of Orwell's later political writing. It was inevitably taken up and read as a simple anti-Communist tract during the Cold War, and this was an altogether too simple reading. If the story is read carefully, there is plenty in it to disconcert conservatives—with a large or a small 'c'—since the message is not that the Russian Revolution was a bad thing but that it had gone in the wrong direction; the critique is from the Left. Orwell's theme is, in Trotsky's phrase, the Revolution Betrayed. Some readers did assume that Orwell's attack on Stalinism was written from a Trotskyite position. But in the story, shortly after the Revolution Snowball–Trotsky and Napoleon–Stalin proclaim that supplies of milk and apples are to be kept for themselves (that is, preserving the privileges of the Leninist élite). Snowball becomes a martyr, but his initial behaviour is open to question. If there is a coherent position implied in *Animal Farm* it seems closer to Anarchism than to any point in the spectrum of Socialism and Communism, even though Orwell always regarded himself as a democratic socialist. In fact, he did not have a wholly coherent position, beyond a belief in basic human decency and honesty and a detestation of totalitarianism. Furthermore, the form of the beast-fable would impose its own pattern, blurring too diagrammatic a political meaning. What began as a revelation that a revolution had been betrayed ends with something like the sad sense that most human aspirations tend to go astray.

Nineteen Eighty-Four became equally famous, though a rather less distinguished work. For several decades the date '1984' served as a shorthand term for a grim regimented future; now that it has become just one more date, moving further into the past every year, some of the peculiar power of Orwell's novel has been dissipated. The political ambiguity of the book has long been acknowledged. Is it primarily an attack on Stalinism in particular, or on totalitarianism in general? Does 'IngSoc' represent Orwell's disillusionment with the British Labour Government, as was widely believed in America, or was he, as he claimed, firm in his democratic socialist convictions, and exposing a perversion of the ideal? The book also raises a less explicit problem about the nature of Orwell's socialism. The whole body of his work shows him to have been an essentially retrospective and nostalgic writer. In his essays and autobiographical writings he frequently looks back to a vanished good place, often sentimentalized as a comfortable working-class home before the war. In his last prewar novel *Coming Up for Air* the hero George Bowling is engaged in a doomed semi-Proustian search for the lost scenes of his childhood. In this orientation Orwell resembles many other English writers of his generation who did not share his ideological commitments. There was inevitably a conflict between this temperamental stance and a commitment to socialism, which believes the past to have been a matter of injustice and oppression and is committed to the establishment of a better world in the future. This contradiction in Orwell has been emphasized by his critics on the Left.

It seems to me peculiarly acute in *Nineteen Eighty-Four*. The IngSoc regime rewrites the records of the past in order to keep them in line with the political demands of the present day, as was standard practice under Stalinism. Winston Smith is engaged in this work at the Ministry of Truth, until he revolts against the process and sets out on a lonely and dangerous attempt to discover the real truth about the past. His quest is a re-enactment, on a much larger scale and in a very different context, of George Bowling's. The values of a more human past are partially recoverable in the novel, as in the archaic working-class enclave where old patterns of living are maintained, and in the carefree sex which Winston enjoys for a time with Julia, in defiance of the puritanical norms of IngSoc.

But by the end of the book the tyrannical present has triumphed over the past. Not only are the historical and journalistic records changed, but the official language, Newspeak, is designed so that subversive thoughts cannot even be uttered. The book ends with unbearable defeat and betrayal—finally Winston comes to love Big Brother. Yet that it is not quite the end; there is a faint suggestion of hope in the Appendix on Newspeak. This describes how the literary classics of the past are being rewritten in Newspeak: 'These translations were a slow and difficult business, and it was not expected that they would be finished before the first or second decade of the twenty-first century.' Literature resists the ultimate totalitarian transformation, which suggests some persistence of traditional human values. But it is a frail hope.

It is a limitation of *Ninety Eighty-Four* that it cannot be read out of the context of its origins in the way that *Animal Farm* can. Orwell was wrong about a number of things; language does not work as he thought it did, and Newspeak is an incoherent conception. More importantly, Orwell was wrong about the absolute power of a totalitarian system to change reality, as we have seen from the collapse of global Communism; no doubt he would have rejoiced to have been proved wrong. *Nineteen Eighty-Four* is a compelling narrative, bringing together futurology, the novel of ideas, and political satire. It is also a deeply pessimistic, almost despairing book. These qualities may partly reflect the final illness from which Orwell was already suffering. But they also indicate the anxieties of the time, as we see from the similarly pessimistic tone of *The Age of Longing* by Orwell's friend Koestler. Domestic as well as international politics may have been a factor, for the last years of the Labour Government were accompanied by a weariness of spirit which recalled that of the final months of the war (the Labour leaders were in a physically exhausted state, and some of them died soon after giving up office). Orwell drew on both these phases in his novel. The future often turns out badly, but never badly in the way that futurological fiction anticipates. Such books are, in any event, really and inescapably about the age in which they are written.

III

In contrast to fiction, poetry was at a low ebb in the second half of the 1940s. For two or three years, while the poetry boom of the war years declined, anthologies and collections continued to appear containing work which had been written during the war. The ambitiously titled *New Lyrical Ballads* (1945) was an anthology intended to give a poetic voice to the Left. The contents aim at a populist simplicity, using ballad forms and folksong, evoking working-class life and English radical traditions, attacking fascism and capitalism, commemorating battles from the point of view of the common soldier, and expressing admiration for the Soviet Union and its war effort. The poetic level is not high; the established poets found it difficult to write at the proper level of simple directness without being self-conscious about it, whilst the amateur poets were merely incompetent. One misses the sardonic note that Brecht would have given to such songs and ballads. But Randall Swingler's 'Sixty Cubic Feet' works well. It begins:

> He was the fourth his mother bore
> The room was ten by twelve
> His share was sixty cubic feet
> In which to build himself.

The ballad goes on to tell how the child has sixty cubic feet of space at school in a class of forty, and then, when he has gone to work in a mine at the age of fourteen, 'in sixty feet of dust and gas | He lay and hacked the coal'. He is called up to the army, and given the same amount of space in a tent on a wet hillside. He falls ill and dies:

They brought him from the hospital
    They brought him home alone
In sixty cubic feet of deal
    That he could call his own.

The work of poets who had served in the Middle East was collected in *Middle East Anthology* (1946), while *Personal Landscape: An Anthology of Exile* (1945) reprinted material from that magazine. Several of the poets who had been in Egypt brought out collections after the war, such as Lawrence Durrell's *Cities, Plains and People* (1946) and *On Seeming to Presume* (1948), G. S. Fraser's *The Traveller Has Regrets* (1948), Bernard Spencer's *Aegean Islands* (1946), and Terence Tiller's *Unarm, Eros* (1947). Keith Douglas's *Alamein to Zem Zem* was published posthumously in 1946, containing a selection of poems as well as the prose narrative, though his *Collected Poems* did not appear until 1951.

In 1945 a first collection by a young poet, Philip Larkin's *The North Ship*, came out and aroused no interest. Its significance now looks more historical than intrinsic, as Larkin himself acknowledged in his preface to the 1966 edition. This very early work shows its origins in wartime Neo-Romanticism; the verse is that of a careful craftsman, but it is dominated by Yeatsian rhetoric, and does not suggest Larkin's mature achievement. That emerged, as he describes in his preface, when he abandoned Yeats as a master for Thomas Hardy. *The North Ship* was quickly followed by two novels, *Jill* (1946) and *A Girl in Winter* (1947); anyone at the time who read the three books might reasonably have assumed that Larkin's real gifts were for fiction and that his poetry was not much more than immature apprentice work. *Jill* is a remarkably accomplished novel for a writer in his early twenties. It clearly draws on Larkin's own experiences as a gauche provincial scholarship boy, but it is not autobiographical in any very obvious sense. It is set in Oxford in the autumn of 1940, against a background of growing shortages, young men going off to the war, and the bombing of London and other cities. John Kemp, just arrived in an Oxford college from a working-class home in the North of England, is bemused by Christopher, the self-assured freshman from a minor public school with whom he has to share rooms, and Christopher's would-be smart friends. In order to preserve his psychic balance John invents a sensitive and thoroughly middle-class young sister called Jill, with whom he maintains a fictitious correspondence. The story gets complicated when he discovers a seeming embodiment of Jill in a 15-year-old schoolgirl whom he keeps running into, and who turns out to be the cousin of Christopher's girl-friend.

Jill works on several levels. It provides a compelling narrative, with poetic pasages that anticipate the later Larkin, and there is one section of great power in which John returns to his home city after it has been heavily bombed, looking for his parents. (Here Larkin drew on his memories of the blitz on Coventry, where he grew up.) In its acute awareness of the codes and markers of class difference *Jill* anticipates the fiction and drama of the 1950s, whilst describing, in the antics of Christopher and his friends, the last feeble flickers of the prewar glamorous Oxford. The principal difference between *Jill* and later books is that John regards the upper classes with helpless fascination rather than anger or derision. *A Girl in Winter* also has a lonely central figure in a wartime setting, Katherine Lind, a refugee from an unnamed Continental country who is working as an assistant in a provincial public library. The story is divided between a dismal winter's day in the library and Katherine's memories of a sunny prewar holiday that she spent at the age of 16 with a comfortably off English family at their house in the Cotswolds. She has written to them to re-establish contact and is expecting a visit from the son, now in the army. Larkin wrote *A Girl in Winter* as a novel of deliberate aesthetic concentration in the manner of Virginia Woolf. To some extent it is, but his real achievement is the traditional one of creating character; the lonely but self-contained Katherine comes across convincingly, as do the other characters, notably her bullying but pathetic boss, Mr Anstey.

After *A Girl in Winter* Larkin attempted another novel, but could not finish it and abandoned fiction, despite his

exceptionally promising start. The second novel's conscious
formality and aesthetic saturation may offer a clue to his
inability to go on; whatever can be achieved in the 'poetic'
novel might be done with still more concentration in verse.
Another factor may lie in Larkin's personal development;
neither novel is autobiographical, but in the fictional creation
and exploration of young and solitary consciousnesses Larkin
was perhaps doing something to face and define his own
emotional state, and the process could not continue inde-
finitely. Many writers write poetry to this end at the beginning
of their careers and then go on to become novelists. Larkin
reversed the order. Nevertheless, the novelist in him left his
mark in the interest in character and the rendering of social
surfaces that we find in the later poetry.

Another gifted poet who appeared at this time was R. S.
Thomas. He was a Welsh clergyman who had published poems
in *Horizon* and *Poetry London* but lived reclusively in rural
Wales; his first book, *The Stones of the Field*, was brought out
by a small Welsh publisher in 1946 and took some time to
arouse any interest. Nevertheless, it marked the start of his
career as a poet of high integrity and indifference to fashion.
To say that Thomas writes about the countryside and its
inhabitants is true, but could be misleading. The poet he most
recalls is his earlier namesake, Edward Thomas, but R. S.
Thomas writes about a harsher landscape—the Welsh hill
country—and his attitude to its inhabitants is not so gentle.
There is a sense in which he writes in the spirit of Wordsworth,
not the Wordsworth whose heart was gladdened by daffodils
but the poet who was moved by the sufferings of the rural
poor and who found in aged countrymen objects for wonder-
ing contemplation. Thomas responds to the Welsh 'peasants',
whose lives seem shrunk by hard and bitter work to a core of
taciturn endurance, with both admiration and exasperation,
and a Christian desire to find an immortal spark within them.
In his poem 'A Peasant' Thomas introduces one such figure
under the name of 'Iago Prytherch', who recurs in subsequent
poems:

> Yet this is your prototype, who season by season
> Against siege of rain and the wind's attrition
> Preserves his stock, an impregnable fortress
> Not to be stormed even in death's confusion.

Though Thomas writes in English and learnt Welsh only in later life, he is a Welsh nationalist with a strong though bitter attachment to the soil of his country. He is painfully conscious of the weight of Welsh history and myth, and in one poem describes Wales as a country with no future and no present, only a past. Earth is a key element in his poetry, whether as native soil, or as the fundamental factor in rural existence, both the source of all new life and the element of death, the grave. Thomas's poems offer variations on a limited number of themes, but he works within the limitations with fine poetic art. His poems combine imagistic observation with epigrammatic sharpness and occasional moments of poignant feeling. *The Stones of the Field* was followed in 1952 by another aptly titled collection, *An Acre of Land*, though Thomas did not receive proper recognition until a selection from both these books, with some later work, came out from a London publisher with an introduction by John Betjeman in 1955.

In 1946 Thomas's celebrated namesake and compatriot, Dylan Thomas, published *Deaths and Entrances*, his first collection for several years, which brought together the poems he had written during the war and just after it, including some which became famous for their lyrical fusion of the autobiographical and the pastoral, such as 'Poem in October' and 'Fern Hill'. At the end of the 1940s two poets whose names had been closely associated with Dylan Thomas's, though not always to their advantage, made significant developments. George Barker was born a year before Thomas, and their careers developed in parallel in the 1930s; both poets had left school at the age of 16 and were self-mythologizers with a Celtic aura. Barker grew up in London but he had an Irish mother, whom he celebrated in one of his best and best-known poems, the tender but rattlingly punning sonnet 'To My Mother'. Barker, like Thomas, was fascinated by words, but he lacked Thomas's concentrated experimental skill; he was a more slapdash and cheerful writer, given to noisy sound effects and exuberant word-play, with an uncontrolled love of puns. But he also was more of an intellectual than Thomas, and ideas get into his poems in a way that they seldom do in the Welshman's. Barker's collected poetry is desperately uneven but it deserves serious attention. A major element in his poetic persona is his abandoned but still deeply felt Catholicism, which gives his writing about sex a profounder dimension than Thomas's visceral stirrings. A sense of sin and guilt is dominant in the two remarkable works that Barker published in 1950, a long autobiographical poem, *The True Confession of George Barker* and a short novel, *The Dead Seagull*. The poem is reminiscent of Auden's 'Letter to Lord Byron' or MacNeice's *Autumn Journal*, written in a stanza derived from Villon's Testament—'The Frenchman really had the trick | Of figure skating in this stanza'—though in places Barker seems to be driving a vehicle with shaky suspension. The *Confession* is witty, ingenious and often funny, but it continually returns to what the poet regards as fundamental:

> I entertained the Marxian whore—
>   I am concerned with economics,
> And naturally felt that more
>   Thought should be given to our stomachs.
> But when I let my fancy dwell
>   On anything below the heart,
> I found my thoughts, and hands as well,
>   Resting upon some private part.

Villon may have suggested the poem's form, but its spirit is close to that of Baudelaire, to whom Barker had dedicated a sonnet. Baudelaire, in a heterodox version of Catholicism, thought that human love is in its very nature fallen and sinful. Barker takes the idea further, to the point of actual Manichaenism: sex is the work of the devil, and reproduction is inherently evil. Yet man—and specifically the male animal—is trapped inescapably by sex and procreation. Sexual passion is the 'terrible aboriginal calamity' in which Newman, in a famous passage, saw humanity as having been involved. These ideas are dominant in *The Dead Seagull*, a strange and disturbing work, which I find more striking than much of Barker's verse. It tells of a young writer and his pregnant wife who are living idyllically in a cottage by the sea, until a friend of the wife arrives, a predatory blonde *femme fatale* who seduces the young man and destroys the marriage. It ends with the death of both baby and wife in childbirth. The symbolic implications are unmistakable; both the novel and the poem are obsessed with the destructiveness of sex. These ideas have surfaced at intervals in the history of Western culture, though it is surprising to find them given such vehement literary expression in dim postwar England; they belong more naturally to the last *fin de siècle*, when Baudelaire was regarded as an antinomian moralist and Gnosticism was a fashionable cult. They will seem repugnant or absurd to contemporary humanists and liberal Christians alike, and Barker's operatic extremity of feeling does sometimes touch on the ludicrous. Yet the point is not the nature of the ideas, but the dramatic and poetic power with which he expressed them. We see something similar in a much greater poet, Yeats, who was also involved in a bizarre mythology.

W. S. Graham was a poet quietly and intensely preoccupied with language and problems of expression, not with adopting any general stance to the world. He was a working-class Scot who began publishing poetry in the 1940s and whose early work derives from Dylan Thomas at his most opaque and riddling. Graham was seriously pursuing the extent to which poetic discourse could concentrate on the associative aspects of language to the virtual exclusion of the denotative ones; the result tended to be pastiche of early Thomas, of even greater obscurity. But *The White Threshold*, which Graham published in 1949, is a work of notable originality; though still often difficult, it conveys the feeling of a world beyond the words, and an inhabited world. The 'white threshold' of the title is the sea, which is a major theme of his poetry, just as it had been for Dylan Thomas. But whereas for Thomas the sea exists in complex metaphorical relations with other kinds of fluid, such as blood, semen, and sap, for Graham it is simply *there*, the vast alien element on which sailors and fishermen depend for their livelihood, and which is always ready to take their lives:

> Last gale washed five into the bay's stretched arms,
> Four drowned men and a boy drowned into shelter.
> The stones roll out to shelter in the sea.

Graham's finest and fullest treatment of man's dealings with the sea came later, in the long title-poem of *The Nightfishing*, which he published in 1955. This is a splendid work, where the struggle with language enacts the hard human struggle against the elements, and is one of the few major poems of the postwar years.

Those years were often seen, at the time, as witnessing a major revival of poetic drama. It is true that poets, inspired by Eliot's *Murder in the Cathedral* and the Auden-Isherwood plays of the 1930s, tended to try their hand at verse drama, just as Victorian poets had done. These plays were sometimes produced in church halls and by amateur groups, but they had no lasting effect in, or on, the English theatre. The conviction that there was a revival came from the work in the late Forties of T. S. Eliot and Christopher Fry. From the early 1920s Eliot had been exercised by the problems and possibilities of poetic drama, which he saw as a way of returning poetry to a wider audience in the theatre. At that time Eliot regarded 'theatre' as extending far beyond the well-made West End play, which he despised, to take in such popular forms as the music hall and the minstrel show. He drew on them in *Sweeney Agonistes*, his first attempt at writing for the stage. Here he presents raffish modern Londoners in a jazzy, formally inventive verse that includes ragtime songs, but there are deep and complex emotions beneath the brash surface of the verse. Eliot was unable to finish *Sweeney Agonistes*, probably because of his abiding difficulty in projecting *action* as opposed to exploring consciousness. But it was an exciting and promising start, where he goes some way towards illustrating the argument of his critical writings that in a true poetic drama the poetry and the action work and advance together, in a 'ballet of words'. After a long pause Eliot returned to playwriting in the 1930s with *Murder in the Cathedral*, a ritual drama which conveys a Christian historical subject in Aeschylean form; it contains some striking poetry, mostly in the choruses, though nothing actually happens, beyond Becket's pre-ordained death. Nevertheless, it was a great success, and remains the most enduringly popular verse drama in English since the seventeenth century. In *The Family Reunion* Eliot used a modern setting but kept the spirit of a ritual drama. Significantly, though, he adopted the form of the well-made play that he had once dismissed; and the poetic aspect was more subdued than in *Murder in the Cathedral*.

Eliot returned to drama after the war with *The Cocktail Party*, which was first performed at the Edinburgh Festival in 1949. It was very well received, and successfully ran in London and New York; the play still has its admirers. One can say in its favour that for the first time in Eliot's dramatic career, things actually change and develop: Lavinia Chamberlayne leaves Edward and then returns to him, and both come to face and understand their basic mediocrity; Edward's former lover, Celia, gives up her life of cocktail parties and casual adultery, becomes a nursing missionary, and endures a cruel martyrdom. There is some agreeable comedy in the play, particularly in the exchanges between the elderly gossips Julia and Alex, and the presence of moral, philosophical, and religious profundities is lightly signalled. There are suggestions of a mythic dimension beneath the naturalistic surface, notably in the distinguished psychiatrist Sir Henry Harcourt-Reilly, whom Eliot adapted from the miracle-working Hercules in Euripides' *Alcestis*. Yet Eliot had abandoned his earlier ideas about poetic drama, and most of his previous practice, without gaining much in return. *The Cocktail Party* takes over the well-worn mode of modern stage comedy which he had once dismissed, writing in a blank verse that is so unobtrusive as to be indistinguishable from fairly formal prose. There are some vestigial suggestions of ritual, but no ballet of words. Nevertheless, the play was much admired on its first appearance and for some time afterwards, and was regarded as providing one half of the great revival of poetic drama.

Christopher Fry, who represented the other half, had come to writing plays from acting and theatre management, and had begun with religious drama, inspired by the example of *Murder in the Cathedral*. In contrast to Eliot, Fry offered fireworks and colour. His verse plays, *A Phoenix Too Frequent* and *The Lady's Not for Burning*, were huge successes on the West End stage and for some years Fry was thought of as one of the leading writers of his generation. What his plays offered, above all, was richness and exuberance of language, but without any challenging newness of dramatic form. It was the right formula to appeal to weary audiences as the dangers of wartime were replaced by the seemingly unending privations of peace. Fry also went in for exotic settings, such as Ancient Greece in *A Phoenix Too Frequent* and medieval England in *The Lady's Not for Burning*; but neither the places nor the supercharged language did more than thinly disguise the basic nature of his plays, which has been well defined by A. T. Tolley:

Fry's comedies owe a lot to Shaw in the use of false argument that rides on the figurative features of the dialogue. The form of the play, as often with Shaw, is an anachronistic drawing-room drama, with an unusual setting, and most of the 'action' is in the form of conversation. The attitudes of the characters are entirely mid-twentieth-century British middle class; and a great deal of the comedy comes from the dissonance between these attitudes and the setting.[1]

To which one need only add that Fry's 'poetic' flights, so admired at the time, are static arias that neither reveal character nor forward the action. His reputation declined as rapidly as it rose, though the title of his best-known play was sufficiently remembered to be adapted by one of Margaret Thatcher's speech-writers to 'The Lady's Not For Turning'. The vogue of poetic drama in the late 1940s did something to meet a cultural craving of the time, but it offered little real sustenance, being a red herring as well as a blind alley. The dramatic work that within a few years was to have a truly transforming effect was just below the horizon, in the plays of Samuel Beckett and

Bertold Brecht. (Brecht was not totally unknown in England, for Auden and Isherwood had come across his work in pre-Hitler Berlin, but it made no impact there until the mid-1950s.) Later developments showed that drama could be 'poetic' without being written in verse. If poetry is a matter of sensitivity to the sounds and implications, the interactions and resonances of words—and also of the silences between them—then Harold Pinter, who writes in 'prose' is a more poetic dramatist than Fry.

IV

Historically, the period I have been discussing was a time of exhaustion leading to eventual slow recovery. At the General Election in 1950 Labour returned to office, but with a tiny majority; later in the year the modest prospects of increasing national prosperity were overturned by the outbreak of the Korean War, which heightened international tension and led to increased defence spending, precipitating a fresh economic crisis in Britain. At a further General Election in the autumn of 1951 Labour was narrowly defeated by the Conservatives, though this defeat resulted from the vagaries of the electoral system, for Labour had a majority of the popular vote. They kept the loyalty of working-class voters in the industrial areas but had lost much of the idealistic middle-class support they received in 1945. The new Conservative Government, with the ageing Winston Churchill back in office, was studiedly moderate, and made no attempt to overturn the Welfare State and the other popular achievements of the Labour Government.

In 1951, during the last months of Labour rule, a Festival of Britain opened in London on the South Bank of the Thames. It was intended to give the people a taste of colour and excitement amid the shortages and drabness that had prevailed for so long, because of the need to sacrifice domestic consumption to the make-or-break export trade. Among other things, the Festival provided an exciting display of new achievements in science and technology, and of innovations in design and architecture. The latter were in some respects

ahead of their time, and had little immediate influence. The
Festival was enormously popular and genuinely did something
to lift the national spirits—there were lesser festivals up and
down the country. The South Bank site contained one permanent building, the newly built Royal Festival Hall, and
many temporary pavilions and other purpose-built structures,
which were variously elegant, ingenious, or made just for fun.
The incoming Conservative Government, wishing to get rid
of one of Labour's most popular achievements, had them
demolished with indecent haste.

Michael Frayn has written a memorable account of the
symbolic significance of the Festival, and of its summary
removal. He sees it as the final manifestation of the humane
ideals—liberal as much as socialist—which had prompted the
aspirations of ordinary people during the war and which the
Labour Government had struggled to put into practice:

Festival Britain was the Britain of the radical middle-classes—the do-
gooders; the readers of the *News Chronicle*, the *Guardian*, and the
Observer; the signers of petitions; the backbone of the BBC. In short,
the Herbivores, or gentle ruminants, who look out from the lush
pastures which are their natural station in life with their eyes full
of sorrow for less fortunate creatures, guiltily conscious of their
advantages, though not usually ceasing to eat the grass.. And in
making the Festival they earned the contempt of the Carnivores—the
readers of the *Daily Express*; the Evelyn Waughs; the cast of the
Directory of Directors—the members of the upper- and middle-
classes who believe that if God had not wished them to prey on all
smaller and weaker creatures without scruple he would not have
made them as they are [. . . .] for a decade, sanctioned by the exi-
gencies of war and its aftermath, the Herbivores had dominated the
scene. By 1951 the regime which supported them was exhausted, and
the Carnivores were ready to take over. The Festival was the last, and
virtually the posthumous, work of the Herbivore Britain of the BBC
news, the Crown Film Unit, the sweet ration, the Ealing comedies,
Uncle Mac, Sylvia Peters . . . all the great fixed stars by which my
childhood was navigated.[2]

It merely remains to add that the Carnivores of 1951 were mild
creatures compared with their successors of thirty years later.

5

Sequences

A characteristic of the English novel in the late 1940s and 1950s was the appearance of works in multiple parts: trilogies or quartets, or *romans fleuves* running into many volumes. Writers who had given up several years of their life to the war found, once they had time to write again, that they had a lot of accumulated experience to unload and put in perspective. Some of them looked back across the wide gulf of the war to childhood, adolescence, and early manhood in what memory often presented as sunnier and more settled times. A retrospective mode was common, where the recovery of the recent or more remote past seemed a necessary part of making sense of the present, in a loose analogy with psychoanalysis. Such a process involved writing at greater length than the conventional novel of eighty or a hundred thousand words. Anthony Powell remarked, 'After the war when I came out of the army and returned to the writing of novels, I decided that the thing to do was to produce a really large work about all the things I was interested in—the whole of one's life, in fact.'[1]

These explorations of the past were already under way during the war, when Joyce Cary made an extensive *recherche* of the earlier decades of the century in his Jimson-Wilcox-Munday trilogy. L. P. Hartley looked back to Edwardian England in *The Shrimp and the Anemone* (1944), which was followed by *The Sixth Heaven* (1946) and *Eustace and Hilda* (1947); the last novel gave its name to the whole trilogy when it was published in one volume. Hartley was born in 1895, and his sequence deals with periods that he himself lived through, beginning with the childhood of his central figure Eustace Cherrington before the First World War, bracketing off the war itself, then picking up Eustace's life as a young man at

Oxford just after it. Hartley was an intensely Jamesian writer, and James's influence is apparent throughout: in his care for the precise rightness of fictional form; his interest in childhood and the ways in which its innocence is upset by the adult world; and in the use of luminous symbols. The first volume opens with a dazzling description of an incident in a rock-pool, in which a shrimp is partly swallowed by a sea-anemone, to the perturbation of the young Eustace, who tries to separate them and so ensures the demise of both. The episode sends symbolic resonances through the many pages that follow it: the shrimp stands for Eustace who is a mild and passive boy, though with a rich fantasy life, while the anemone symbolizes his good-hearted but domineering elder sister, Hilda.

The vividly rendered periods and settings of the trilogy gave it a strong though superficial nostalgic appeal when it first appeared, which has continued: an Edwardian seaside resort, Oxford in the early Twenties (in a soberer version than Evelyn Waugh's), a summer in Venice. But if *Eustace and Hilda: A Trilogy* is, as I believe, some kind of a masterpiece, then the reason goes deeper than mere nostalgia. Hartley, like James, understood the human capacity for hope and disappointment and sadness, and enacted these qualities in formally beautiful narratives. He creates memorable characters, and he adds to the familiar modes of English social comedy a note of the uncanny and sinister, as when Eustace, in the last volume, encounters an everyday-looking ghost in a Venetian palazzo. Above all, there is Hartley's wonderfully empathetic presentation of childhood in the character of Eustace. Walter Allen has given an admirable account of this achievement:

Eustace seems to me among the most completely successful character-creations in contemporary fiction, and, at first glance, everything might seem to have been against Hartley here. A nervous, diffident, timid, over-scrupulous, as it were professionally delicate little boy who grows into a delicate, unsure, charming young man with literary ambitions: one of the stock characters of English fiction since Butler wrote *The Way of All Flesh*, we might think. Yet Eustace is a fresh, unique being both as a small boy and as a young man at Oxford and in Venice. He is so because of the special quality of his mind . . . The plot enhances him, elevates him to his full significance. At the same time, by striving to squeeze from the situation the last drop of drama inherent in it, Hartley has been repaid . . . by beauty of form, the rarest quality in English fiction. Eustace partakes of that transfiguring quality.[2]

One can add to Allen's judgement that Hartley has also created splendidly convincing minor figures, such as Jasper Bentwich, the irascible, cynical English expatriate who befriends Eustace in Venice.

Hartley's most celebrated and popular novel is *The Go-Between* (1953), whose opening sentence has passed into proverbial wisdom: 'The past is a foreign country; they do things differently there.' This points to the strong nostalgic appeal of its uncovering of the hidden dramas of life in an Edwardian country house. The theme is very Jamesian: a small boy is traumatically caught up in the intrigues of adults, and as he looks back on the experience from middle age we realize that it has left him a psychological cripple. The *Go-Between* is very beautiful and very painful, though narrower in scope than the trilogy. I think it may be the most perfectly realized of modern English novels, though to say that is not to bestow unlimited praise; perfection is not, paradoxically, the greatest quality a novel can have.

There is a comparable feeling for childhood in Jocelyn Brooke's *The Orchid Trilogy*, whose constituent volumes were first published as *The Military Orchid* (1948), *A Mine of Serpents* (1949), and *The Goose Cathedral* (1950). Brooke is one of the few authors of the retrospective sequences of those years to be openly influenced by Proust; for the most part his influence, though difficult to escape altogether, was something to resist rather than succumb to. *The Orchid Trilogy* is a work of fictionalized autobiography, in which the character of 'Jocelyn' bears much the same relation to the author as 'Marcel' does to Proust. We first encounter 'Jocelyn' as a very unusual small boy with an informed passion for orchids and fireworks. Scenes of childhood in rural Kent are contrasted with those of the author's adult life when he is serving in the

Royal Army Medical Corps during the war. The fluidity of
time and memory is Proustian, and so is the use of certain
objects which recur in the narrative, such as a distant water-
tower, and the 'goose cathedral', which is the boy's fanciful
name for a boat-house. The delicate sensibility is stiffened
with wit and humour, and the work is often very funny.
Ronald Firbank is another influence, evident in Brooke's
conversational style, while there is something traditionally
English in his fascinated interest in eccentric types (including,
perhaps, himself), reminiscent of Aubrey's *Brief Lives*.
Brooke was primarily writing for his own pleasure, but *The
Orchid Trilogy* is a work of genuine originality—one can apply
to it that over-used reviewers' word, 'quirky'—and very great
charm.

   In Chapter 2 I described Evelyn Waugh's first sustained
work of retrospective fiction, *Brideshead Revisited*. This was
written during the war, when the author was recovering from a
broken leg sustained during his first practice parachute jump,
and, as Waugh later acknowledged, was a work of nostalgic
fantasy, written—and gratefully received—as a compensatory
escape from the privations of the time. Over the years Waugh
was able to regard his wartime experiences with greater detachment, and he returned to them in *Men at Arms* (1952) and
its sequel, *Officers and Gentlemen* (1955), which are written in
a cooler, more economical prose. These two novels described
the wartime adventures of Guy Crouchback, a mild, melancholy Catholic gentleman of private means. At the outbreak of
war, when *Men at Arms* opens, he is 35 and living quietly in
Italy, as he has done since his wife left him several years
before. He feels himself to be at odds with the modern world,
and rejoices in the Soviet–German pact of August 1939 for
bringing together its most detestable manifestations, Nazism
and Communism. 'The enemy at last was plain in view, huge
and hateful, all disguise cast off. It was the Modern Age in
arms. Whatever the outcome there was a place for him in that
battle.' Guy returns to England to join the army, and this is
something he finds remarkably difficult to achieve during the
Phoney War. Eventually he gets into an ancient and idio-

syncratic regiment called the Halberdiers. His military career proves inglorious: he takes part in an attempt to occupy a Vichy French colony in West Africa, which is a total failure, and in the disastrous campaign in Crete in 1941, from which he barely escapes with his life.

At first it was not altogether clear what the two books added up to, and the second of them seemed to end inconclusively. There was, though, a welcome return of Waugh's comic spirit, most fully evident in three finely grotesque characters, who are far from reflecting Guy's romantic ideals of soldierly life: Captain Apthorpe, Brigadier Ritchie-Hook, and Corporal-Major (later Major) Ludovic. Then in 1961 Waugh published *Unconditional* *Surrender* (called *The End of the Battle* in America), which was described as the final volume of a trilogy about Guy Crouchback. This, it seems, is what Waugh always intended to write, but he had found himself blocked for some years after the second volume. As well as being a distinguished novel in its own right, *Unconditional Surrender* made more sense of the two previous volumes and changed one's understanding and valuation of things in them. In 1965, not long before he died, Waugh put the three together, with some revisions, to form a single long novel bearing the title *Sword of Honour*. (It is still common to refer to it as a 'trilogy', though the final version does not indicate its tripartite origin; it reads as a unified work of nearly 800 pages in eleven chapters.) *Sword of Honour* is generally and rightly regarded as the finest English novel to come out of the Second World War. It suggests in more than one way the great works of the nineteenth century: the comic flights are reminiscent of Dickens and can stand to the comparison, whilst the relentless moral exploration—and exposure—of Guy's romantic delusions has a Stendhalian sharpness. Guy embodies Waugh's myth of the Gentleman—Christian, conservative, soldierly, of an ancient family—that had detrimentally dominated *Brideshead Revisited*. As *Sword of Honour* progresses we see the myth falling apart. The traditional military ideal is undermined by the buffoons and knaves in uniform who abound during the war; from an ally of Germany, Soviet Russia becomes an ally of Britain; Guy acquires no trace of glory in battle, and learns bitterly and reluctantly how the demands of *Realpolitik* systematically betray the ideals .with which he had embarked on the war. 'This is not soldiering as I was taught it' he remarks towards the end of the war, when forced to acquiesce in the liquidation of a Yugoslav royalist by the Communists. Later, when Guy is serving with the partisans in Yugoslavia, a Hungarian Jewish refugee woman talks to him about the way many people had hoped for war to attain national or political ends, or to achieve a form of personal redemption: 'Even good men thought their private honour would be satisfied by war. They could assert their manhood by killing and being killed. They would accept hardships in recompense for having been selfish and lazy.' Guy sadly acknowledges that this had been his own attitude: '"God forgive me," said Guy. "I was one of them."' His personal mythology has collapsed. Not long afterwards the woman, having escaped the Nazis, disappears into Communist captivity.

During the novel Guy also learns from his devout but down-to-earth father to see his Catholicism in a less romantic light; a lesson reinforced when he goes to confession in Egypt to an Alsatian priest who proves to be an enemy spy. Guy survives the war to lead the life of an English country gentleman, having learnt much. His wife, with whom he had been briefly reunited, is killed by a flying bomb in 1944, having just had a baby fathered by the odious Trimmer, a young officer who had once been a hairdresser and has been turned into a war hero, on trumped-up grounds, by the government propaganda machine. Guy adopts this boy as his own; after the war he marries a sensible Catholic woman but has no more children, so the son of Trimmer is the only heir to Guy's name and estate. (Originally, in *Unconditional Surrender* Guy is described as having two sons of his own, but Waugh took them out in the revised version; Guy's personal childlessness is more thematically apt.) We may see here a cynical embodiment of the reconciliation between classes which was a popular ideal during the war; the real point for Guy, though, is that the child is being brought up as a Catholic, and blood and heredity are of no importance compared with the absolute worth of every individual soul.

There is, it is true, a note of benign cynicism about the end of *Sword of Honour*, but here and elsewhere the book offers no sure foothold for the reader's complacent assumptions. It is, I think, part of its quality that attitudes implied in the text are regularly undermined or otherwise called into question. Waugh continued to write and talk like a reactionary and a snob and a figure of clownish malevolence, though these stances were all a matter of self-dramatizing projection. In *Sword of Honour* we see him writing as an imaginative artist rather than a public persona, distancing himself from Guy's experiences and beliefs, even though they were close to what we know to be Waugh's own (Waugh and Guy had much the same war). The capacity to regard his own intensest experiences with analytical detachment is evident in a brilliant short novel, *The Ordeal of Gilbert Pinfold* (1957), which Waugh published between *Officers and Gentlemen* and *Unconditional Surrender*. It makes comic and disturbing fiction out of the painful ordeal of an elderly novelist who goes temporarily mad on a voyage to the Far East; everything that happens to Pinfold, it seems, had happened to Waugh himself. Waugh's capacity for detachment and for making use of any experience, however distressing, is the sign of a major artist.

During the late 1950s the successive volumes of Lawrence Durrell's *Alexandria Quartet* rapidly acquired both a wide and eager readership and, for a time, high critical esteem. They were *Justine* (1957), *Balthazar* (1958), *Mountolive* (1958) and *Clea* (1960); in 1968 they were brought together in a single volume as the *Quartet*. Durrell was already known as a talented poet and a writer of books about Mediterranean islands; and, to some, as the author of a prewar novel, *The Black Book* (heavily influenced by Henry Miller) whose obscenity meant that it had to be published in Paris. The Alexandrian novels brought Durrell a much larger audience and made him a best-seller; they were immensely successful in France. They provided multiple forms of exoticism. First, there was the presentation of the ancient city of Alexandria, a melting-pot of races, cultures, religions, and languages; part of Africa yet also an outpost of Mediterranean Europe. Then there was the overblown, lush quality of Durrell's prose. He described a colourful, rather sinister, cosmopolitan way of life, with much sexual variety and occasional displays of violence (the volumes bore epigraphs from Sade): it was all glamorously different from the drab, sober existence of England in the Fifties, slowly recovering from the exhaustion of war. Furthermore, the *Quartet* offered itself as a work of ambitiously modernist fiction, in contrast to the glum realism of the new young novelists and kitchen-sink dramatists. It was embraced by critics and readers as the latest manifestation of the tuppence-coloured, in its perennial struggle with the penny-plain in English culture.

Durrell took his own work very seriously, invoking Proust and Joyce, and Einstein's relativity theory, to explain his treatment of time. In fact, the form of the sequence is not particularly complex or demanding; the first three volumes cover from different angles the connected lives of a group of Europeans and Egyptians during a few years in the late 1930s, with an excursus into an earlier period in *Mountolive*; in *Clea* the story moves forward into the war years. It is not difficult to see why the *Quartet* was a success at the time (its popularity in France may be partly due to Durrell's taste for gallic-style aphorisms about love and art, which sound better in French than in English). While the work's early fame is understandable, the subsequent decline of its reputation is equally so. Though it contains some marked local successes, the *Quartet* as a whole now appears as ramshackle and pretentious. Durrell's prose, often praised for its richness, sometimes fails badly: a phrase such as 'the sad velvet broth of the canal', for instance, is inept, since the implications of the words neutralize each other. Justine's husband, the Coptic million-aire Nessim, has a fine car which often appears in the narra-tive; early on it is referred to as his 'great silver Rolls', and thereafter it is the 'great car' or the 'great limousine'. Durrell freely applies the inert 'great' when he wants to raise the tone of the narrative. Invoking the glamour of high life he drops into pastiche of F. Scott Fitzgerald: 'the long cool balconies of the summer residence echoed night after night to the clink of ice in tall glasses'; 'the foundering plunge of saxophones crying to the night like cuckolds'.

Durrell is much better at writing scenes of collective action: a duck-shoot at dawn, a nocturnal fishing expedition, a bizarre Coptic funeral. Though his central characters remain basically uninteresting—and this, perhaps, is the major flaw of the sequence—some of the lesser ones are memorable, like the comic figure of the elderly transvestite English police-officer, Scobie; or Nessim's farmer-brother, Narouz, an embodiment of Byronic power and savagery. The *Quartet* aspires to modernistic self-reflectiveness about writing fiction: two of its characters are novelists, as was Justine's first husband, dead before the story opens, who had written a book about her. Darley, who narrates parts of the *Quartet*, is a minor novelist, while Pursewarden is supposed to be a major one. Darley has the same initials as Durrell, who has admitted to a partial identification with both characters: 'I believe I am both Pursewarden and Darley, but intermittently in flashes; but different from either.'[3] Pursewarden is treated with excessive reverence in the *Quartet*, and far too much space is given over to his aphorisms and reflections, which are mostly banal, and sometimes repulsive, as when he remarks, 'In my country almost all the really delicious things you can do to a woman are criminal offences, grounds for divorce.'

There is a recurring note of cruelty, and Durrell's characters suffer in painful ways. By the end of the work several of them have died, and those that remain have nearly all been maimed or disfigured. Some reviewers recognized at the time that the *Quartet* was an embodiment of the Romantic Agony, recalling the Decadence of the late nineteenth century; that, certainly, is the strongest impression it now leaves. Nostalgia was a dominant force in postwar English writing; but Durrell's sequence looks back to a very remote time and culture; much more remote, in fact, than its ostensible period. He had been born in India and lived nearly all his adult life out of England, a country he knew little of, disliking what he did know. Martin

Green, in a hostile essay, has remarked, 'As a result of living so long out of England, and so much among embassy officials, Durrell's mind is a museum-piece.'[4] The strong immediate appeal of *The Alexandria Quartet* lay in its exoticism and its difference from the drab quality of life of England in the 1950s. But in the longer run the exoticism faded, and the sequence remains an ambitious and isolated oddity.

C. P. Snow's *Strangers and Brothers* was another popular sequence in the 1950s, longer, and much longer running. The first volume, also called *Strangers and Brothers*, had appeared in 1940 (later retitled *George Passant*, after its name was applied to the whole work). Snow did not continue with the sequence until after the war, when *The Light and the Dark* came out in 1947; thereafter, volumes appeared at regular intervals until it was concluded with the eleventh, *Last Things*, in 1970. The fact that both Snow and Durrell were so highly regarded tells one something about the range of taste at the time, though it is unlikely that they appealed to quite the same readership. Where Durrell writes about marginalized expatriates in a decadent cosmopolitan milieu, Snow is concerned with the men in high places who make decisions and run things, and his sequence moves through the various sites of power in English society: legal, academic, scientific, administrative, political. We are usually in London or Cambridge. And whereas Durrell's prose was extravagantly colourful, Snow's is sober-suited and workmanlike. Yet both novelists were writing of what they knew.

It was one of Snow's strengths that his career was that of a man of affairs rather than just of letters, so that he knew from the inside much more about how society is actually run than most novelists do. He had begun his career as a physicist, was an academic at Cambridge during the 1930s, an administrator in the civil service during the war and in business after it, and was briefly (and not very successfully) a minister in the Labour government of 1964. The central figure and narrator of the sequence, Lewis Eliot, goes through a similar variety of experiences. We meet him first as a young man in a provincial city in the 1920s. He becomes a lawyer and we follow him to the Inns of Court in London and then to Cambridge, where he takes up a college fellowship. He has a widening circle of acquaintance, including the aristocracy and prominent businessmen, and he enjoys the intimate friendship of a rich Anglo-Jewish family. During the war, as a temporary civil servant, he is involved with the setting up of an atomic research establishment. In later years he is knighted and enjoys the friendship and confidence of the great ones in the land. He has given up active public life and become a writer. Snow did not try to disguise the close identification of himself and his character: 'I would have thought that in depth Lewis Eliot is myself. In a good many of his situations, a good many of his external appearances he is not me, but in any serious and interesting sense he is.'[5] This is a significant admission, given Eliot's high degree of self-regard and self-approbation throughout the sequence.

Snow has bequeathed a phrase to the language, 'the corridors of power', which is the title of one of the later volumes of *Strangers and Brothers*. Certainly, power is what he finds most interesting and writes best about; he is very good at describing the struggles of the committee-room, whether the context is academic, scientific, or governmental. He presents a masculine world, where important decisions are made in front of blazing coal fires, with the curtains drawn against the darkness and the inclement weather outside, and, by implication, to exclude the uninitiated. ('Cosy' is a favourite epithet of Snow's in such situations.) At a time when most modern fiction is concerned with personal life and the emotions and sensibilities of individuals, Snow's informed interest in the public realm and the exercise of power is salutary, and a reason why he appeals to readers who are bored by other novels. But he has a very unbalanced talent, with little capacity to explore persuasively the inner lives and feelings of his characters. Lewis Eliot's attempts to convey the emotional complexities of his two marriages, for instance, easily lapse into sentimentality. In such situations Snow's flat prose is quite inadequate to express any real subtlety of feeling.

One central problem is that Snow treats separately, in

separate volumes, the public and personal strands in the lives of his characters, with a fragmenting effect. Discussing Snow, Stephen Wall has aptly invoked the Wilcoxes in E. M. Forster's *Howards End*, who 'did not make the mistake of handling public affairs in the bulk, but disposed of them item by item, sharply'. This implies not just a failure of technique, but a deficiency in Snow's sense of human life. The point emerges clearly if one compares his work with that of the novelists he admired, such as Proust or Trollope, with whom he is often compared, although, in Wall's words, he lacks 'Trollope's curiosity, his empathy, his realistic assessment of his own talent'.[6] Snow does, in fact, know that people are ultimately mysterious and contradictory, but he cannot enact this knowledge. *The Masters* (1951), one of his most popular novels, and indeed one of his best, which Lionel Trilling called 'a paradigm of the political life', provides a good illustration. It is concerned with the cabals and intrigues in Eliot's Cambridge college in the year 1937, when the Master is a dying man and the Fellows need to think about his successor. The novel moves steadily through the intricacies of the plot towards its conclusion, and a high degree of suspense is maintained. But Snow says little about the worsening international situation, or the personal lives of Eliot and his friends; as in a detective novel, everything is subordinated to the central plot. At the same time, Snow aims to do more than tell a good tale; he wants to illuminate the human strengths and weaknesses that come out in such a situation. The favoured candidate for the Mastership of Eliot and his party is Dr Paul Jago, who is a very fine human being, as Americans say; a man of imagination and great warmth and indeed nobility of character. We are *told* about these admirable qualities at regular intervals, but we never see them in action. A cipher for a long time, Jago appears by the end of the novel to be an unstable egotist, and the reader may well have come to think that his more prosaic opponent is the better man.

Anthony Powell's *roman fleuve*, *A Dance to the Music of Time*, is one volume longer than Snow's, and is greatly superior as literary art. The first volume, *A Question of Up-*

bringing, came out in 1951, and the twelfth and last, *Hearing Secret Harmonies*, in 1975. Powell's narrator, Nicholas Jenkins, moves through much the same kinds of life as his creator; as a boy at public school in the early 1920s, which is where we first meet him, as an Oxford undergraduate, then as a filmwriter and publisher and frequenter of the more raffish side of London society in the 1930s, and as an army officer in wartime. One can trace the parallels in Powell's subsequent collection of memoirs, *To Keep the Ball Rolling* (1983). Yet it would be misleading to think of *The Music of Time* as an autobiographical novel, as that term is usually understood. Jenkins shows little interest in revealing his inner life and personal development, and he is reticent about his emotions. His gaze is outward, at the people he meets as he goes through the world, with a particular eye for eccentricity and oddity. Powell ministers, above all, to the human love of gossip and anecdote, and *The Music of Time* is a vast collection of interwoven and extending anecdotes. In the 1940s Powell edited John Aubrey's *Brief Lives* and wrote a life of him, and there are evident affinities between the novelist and the seventeenth-century anecdotalist.

The Music of Time is a great work of English social comedy. Inevitably Powell invites comparison with his contemporary and lifelong friend, Evelyn Waugh. They write about similar worlds, and they have a similar comic sense. But the differences are more important, and serve to define the art of each writer. Powell is quite without mythopoeic feelings; there is no equivalent in his work for the myth of the gentleman that captivated Charles Ryder and, for a considerable time, Giles Crouchback until it shattered around him. Nicholas Jenkins appears as an assiduous and fascinated observer of the human spectacle with no general ideas about the world at all. As far as he is concerned, life is simply one damned thing after another, and one has to make sense of them as best one can, enjoy the spectacle and maintain such equilibrium as one can in an unstable world. There is no point in nostalgia for a vanished and possibly better order of things. Jenkins's style is an important element in his attempt to maintain a minimal poise. The initial impression it makes on most readers is that it is remarkably leisurely and intricate and even long-winded; sometimes it puts them off so that they abandon the book. However, one soon gets used to it if one reads on, and it becomes clear that Jenkins's manner is part of his tentacular endeavour, hesitant but persistent, to make some kind of sense of what he undergoes and recalls.

Powell has also been compared with Proust; both writers were concerned with recovering a large tract of past time in a lengthy chronicle, and both were fascinated by extravagances of character. But the resemblances are only superficial. Nicholas is not introspective, and though paintings are an important structural element in Powell's sequence, from the Poussin described in the first paragraph onwards, there is no equivalent to Proust's belief in the transforming power of art. And in his steadily sequential unfolding of the story, Powell treats time in a very traditional way; Jocelyn Brooke, in the fluid chronology of his *Orchid Trilogy*, is much more Proustian. Powell is, in fact, close to the mainstream of nineteenth-century fiction in his form of narrative, and his capacity to create a large gallery of memorable characters, of whom we meet about two hundred in the course of the sequence. Powell's prewar world is, as Walter Allen neatly put it, located where Mayfair meets Soho, and its denizens tend to be minor aristocrats, seldom of the first lustre, and bohemians of all kinds, some of whom are also practising writers, artists, and musicians, together with their wives and girl-friends. There are a few businessmen (some of them shady), rising or falling politicians, and mysterious figures with no visible means of support.

Powell's greatest creation is Kenneth Widmerpool. He dominates the sequence and is one of the most fascinatingly awful characters in modern fiction. We meet him first in *A Question of Upbringing* running out of the mist, a solemn, ungainly, rather absurd schoolboy, but already filled with a ruthless determination to succeed. In later volumes he retains both his absurdity and his ruthlessness, as he rises first in business, then in the army, and after the war in politics and, finally, in academic life. (Widmerpool's progress is like a parody of that of Snow's Lewis Eliot.) He is sometimes humiliated, as when a girl with whom he is in love showers him with sugar at a society dance. But nothing daunts him. Not many people like him, though Jenkins seems to, but he is reluctantly respected. Widmerpool is a superb comic figure, but he also embodies the power of an unsurmountable will, which Jenkins reluctantly admires. Thus, Widmerpool hails a taxi and one instantly appears out of nowhere; Jenkins reflects, 'A cab seemed to rise out of the earth at that moment. Perhaps all action, even summoning a taxi when none is there, is basically a matter of the will.' If there is any underlying thematic element in the sequence, it is Powell's respectful interest in the will. Other characters embody it, too, such as the left-wing literary critic J. G. Quiggin, and the beautiful and ferociously bad-tempered Pamela Flitton, whom Widmerpool eventually marries, and in whom he meets his match. Characters with more desirable qualities, like Nicholas's close friend, Charles Stringham, who is attractive, witty, cultivated, and sensitive, but also aimless and moody, tend to face defeat. He becomes an alcoholic, and during the war dies as a prisoner of the Japanese, a fate in which Widmerpool has indirectly had a hand.

In Nicholas's perception—and Powell's too, no doubt— Stringhams are to be preferred to Widmerpools, but they cannot survive and the future seems to lie with the latter. But this is not a heavily emphasised theme; Powell implies that everybody has something to be said for them, if it is only the quality of making us laugh, and this general charity, traditionally Christian (though religion is conspicuously absent from the sequence), is unusual in twentieth-century literature. Any discussion of *The Music of Time* that does not recognize it as, above all, great comic fiction is liable to miss the point, though it is also true that as the sequence goes on the comedy is shadowed by an elegiac element, as more and more of the characters disappear through death, particularly during the war. I have my reservations about aspects of *The Music of Time*, particularly the power of the central idea of the 'dance' to sustain the whole long structure, and Powell's treatment of time in the last volumes, and I have discussed these elsewhere.[7] But I have no doubt about the quality of his achievement.

Doris Lessing's five-part sequence, *Children of Violence*, is the work of a writer who grew up in what used to be Rhodesia and is now Zimbabwe. She moved to England in 1949 and shortly afterwards published a striking first novel, *The Grass is Singing* (1950), which suggests Lawrence's *Lady Chatterley's Lover* transposed to an African setting. It describes a passionate and doomed affair between a bored white farmer's wife and a black servant. In 1952 Lessing published *Martha Quest*, the first volume of *Children of Violence*; other sections appeared at intervals and it was completed with *The Four-Gated City* (1969). The work is substantially autobiographical, and most of it is set in the narrow, blinkered white community of Rhodesia in the late Thirties and early Forties. Martha Quest is a self-emancipating heroine who anticipates the feminist writing of later decades, notably in her first-hand account of the discomforts of pregnancy. She engages in left-wing politics, starts to educate herself, gets married, becomes pregnant, is divorced, comes to England with her child. Although some of the prose has the lyrical quality that memories of her early African life inspire in Lessing, much of Martha's story is a dull naturalistic plod. But the final volume is very different, in approach and in genre: it is a futurological novel which describes Martha's life as an old woman in the apocalyptic London of the year 2000 (still more than thirty years ahead when *The Four Gated City* appeared). The switch reflects the author's own change of interests and commitment from Marxism and fictional naturalism to Sufism and science fiction.

In 1956 Anthony Burgess's first published novel, *Time for a Tiger*, opened his long career as a prolific and versatile man of letters. It shows the misfortunes of Victor Crabbe, an expatriate English schoolteacher in the racial melting-pot of Malaya during the early 1950s, when the Chinese communists were engaged in a campaign of terrorism and violent subversion. It is evident that British colonial rule is coming to an end, and the process is continued in Burgess's next two novels, *The Enemy in the Blanket* (1958) and *Beds in the East* (1959); in the last of these Victor, having been left by his wife, is accidently drowned just as Malayan independence is proclaimed. The three novels were later brought together as *The Malayan Trilogy* (the American edition had the more elegant and thematically suggestive title, *The Long Day Wanes*, a phrase from Tennyson's 'Ulysses'). The tone of the narration tends to be sour and cynical, and the writing is flat compared with the bravura stylistic displays in Burgess's later novels. There are lively episodes, some of them highly comic, but the work as a whole typifies his perennial tendency to be strong in arresting incidents but weak in fictional structure. The trilogy is one of the first works of fiction to show the winding-up of the imperialism which had been an oppressive presence in such earlier novels as E. M. Forster's *A Passage to India* (1924), George Orwell's *Burmese Days* (1934), and Joyce Cary's *Mister Johnson* (1939). The struggle against imperialism, as it existed in the curious delegated form of the self-governing white colony of Rhodesia, was also a major concern of Doris Lessing's earlier writing. In her work the opposition was clear, literally in black and white. In Burgess's trilogy the coming of independence is seen as inevitable, but the narrator shows open scepticism about it; in Malaya, unlike Africa, there were many racial groups—Malayan, Chinese, Indian, Tamil, Sikh— who, however opposed they may have been to the British, were also opposed to each other.

The last extended work to be looked at here has no obvious relation to the other books discussed in this chapter, or, indeed, to any other English literature of the Forties and Fifties. It stands alone, like so much else by its creator, Wyndham Lewis. He was born in 1882 and began his dual career as writer and painter in the years before 1914; on the eve of the First World War he achieved brief public notoriety as the founder of the Vorticist movement and editor of the loudly iconoclastic magazine *Blast*. His literary reputation is likely to rest on three novels, *Tarr* (1918), *The Revenge for Love* (1937), and *Self Condemned* (1954). But in the 1920s he was engaged on a huge enterprise which was to comprise several related works of fiction and socio-cultural polemics. He was a man equally well endowed with talents, energy, and egotism, who saw it as his lonely mission to scarify the cultural forms, both high and popular, of postwar England. He referred to himself as 'The Enemy' and, in the pursuit of his own ideal of great art, attacked primitivism, democracy, mass civilization, feminism, homosexuality, the cult of youth, and whatever else attracted his ire. These attacks and his pro-fascist stance in the Thirties (he abandoned it in 1939) aroused literary opinion against him. Lewis was often reminiscent of Thomas Carlyle, in his savagely rhetorical onslaughts on what he regarded as the follies of the age. One product of his abandoned large enterprise was *The Childermass* (1928), a book which is difficult both to classify and to read. It is best described as a fantasy of the afterlife, set on a large plain with the Heavenly City in the distance. The principal characters are James Pullman, who is originally described simply as a schoolmaster but who later on appears to be an important writer; and Satterthwaite, usually called 'Satters', a grotesque overgrown schoolboy, who had been Pullman's fag at public school. The dealings between these two are often very funny, for Lewis was a brilliant comic writer; Pullman and Satters provide an example of what Fredric Jameson has called the 'pseudo couple' of opposed but closely related male figures, looking back to Flaubert's Bouvard and Pécuchet, and beyond them to Don Quixote and Sancho Panza, and forward to Beckett's Estragon and Vladimir. Lewis was a master of English prose, whether he was writing comically or descriptively; in the latter vein he was extraordinarily original and powerful, with the painter's eye fully engaged, as we see, for instance, in the opening paragraphs of *The Childermass*.

Yet much of the book is tedious in the extreme, when Lewis uses it as a vehicle for his ideas about the world. The third major character is the Bailiff, a bizarre Punch-like figure who appears to be a senior administrator from the Heavenly City. He serves as the mouthpiece for all the attitudes that Lewis most disliked, a prize representative of the fashionable

exaltation of time and flux. The comic encounters of Pullman
and Satters and the stunning descriptive writing are not
enough to rescue *The Childermass*, but Lewis was very committed to it, and intended to continue it; the first edition
contains an announcement that *The Childermass* was to consist
of three sections, and that the second and third sections would
appear in the autumn of 1928. They did not appear then, nor
for many years.

Eventually, in the early 1950s, the Third Programme of
BBC Radio put on a dramatized version of *The Childermass*.
That was the golden age of radio drama, and the Third Programme was a munificent patron of writers. The producer,
D. G. Bridson, was an admirer of Lewis and arranged a
commission for him to write the subsequent sections of *The
Childermass*, initially for radio transmission and then for
publication in book form. Lewis was then in his seventies and
enduring that most appalling fate for a painter, of going blind.
He was delighted with the commission, which would provide
the financial support to let him complete the work at long last.

The second and third sections were published in 1955 as
Monstre Gai and *Malign Fiesta*, and all three were then reissued together as *The Human Age*. Not surprisingly, the
connections between the first part and the later ones are
somewhat tenuous. Pullman and Satters make their way into
the City, which proves not to be Heaven, but a mundane form
of purgatory called Third City. The Bailiff has diminished
from a powerful ideologue to a kind of gangster. There are air
raids and insurrectionary outbreaks, as Third City is under
attack by Satanic forces. Lewis has constructed his cosmology
with Dante in mind; the plain of *The Childermass* is on the
banks of the Styx, though here Purgatory comes before Hell,
known as Matapolis, the setting of *Malign Fiesta*. Hell, too,
proves to be a rather ordinary big modern city, though hor-
rible atrocities go on in the punishment sections. Both cities
reflect, in their banality and violence, the middle years of
the twentieth century. Lewis decided there were to be four
sections rather than the original three; the final volume was to
be set in Heaven, but he had written only a few introductory

pages when he died in 1957. *Malign Fiesta* ends with Pullman being taken upwards, like Oedipus at Colonus, by two
heavenly messengers.

The successive volumes provide a curious phantasmagoria,
with echoes of Dante and Milton—particularly the war in
heaven—and Swift, plus Arabian Nights magic, with flying
dragons, angels who change their size at will, and nasty devils
who are part-man, part-goat. The narrative is compelling in
places—Lewis is too good a writer for it not to be—but the
writing is often flat and hurried, without the stylistic variety
that was one of the positive features of *The Childermass*.
There is effective local detail, but overall incoherence.

I see Lewis's real and very impressive literary achievement
of the 1950s not in *The Human Age*, but in *Self Condemned*.
This novel is largely based on his experiences during the
Second World War, when he and his wife lived in Canada in
poverty, cut off by exchange regulations from any of his
British income. The central figure, Professor René Harding,
embodies many of Lewis's attitudes and ideas. He is said to
look like the philosopher Descartes, whose Christian name he
bears, and he is a Cartesian dualist, living in a world of
thought and resenting the body and its demands. This dualism
characterized Lewis's own position for many years, as did
Harding's contempt for the common herd, and his misogyny.
Harding is treated with detached irony, and as the novel
unfolds it becomes evident that Lewis is using him as a means
of distancing himself from his earlier ideas and convictions. In
his writing and painting Lewis had long preferred the hard
exterior, the shell or carapace, to the soft interior; in his
literary criticism he had mocked the explorers of consciousness, like Proust and Joyce and Virginia Woolf. But in *Self
Condemned* Lewis accepts interiority; late in the day he
perceives that human beings have depths as well as surfaces.
By the end of the novel Harding has emerged from the miseries
of the hotel room where he and his wife had been marooned
during the long Canadian winters. He is about to move to a
good job in a university in the United States. But his life is
shattered, for his wife has committed suicide, and on the last

page he is significantly described as a 'glacial shell' of a man. In this novel Lewis comes to terms with the humanity that he had so long despised. It gains considerably if one reads it in the context of his other work. But even if one knows nothing about Lewis, *Self Condemned* is a powerful modern tragedy, and one of the outstanding novels of the 1950s.

6

Anger and the Empirical Temper

I

Most of the writers discussed in the last two chapters had begun their careers before 1939. As the Forties turned into the Fifties, critics and general readers were looking for promising young writers but not finding them. A new literary generation was overdue. Angus Wilson had seemed to be the harbinger of one in his short stories that coolly anatomized postwar England, but he was already 36 when his first book appeared. Many young writers born in the early 1920s had served their literary apprenticeship in the armed forces, producing short poems and prose pieces. When peace came it took them several years to establish themselves, to learn to write in more extended and complex forms, and to get a sense of the new world shaping around them.

Meanwhile a climate of expectancy grew up, eagerly and impatiently waiting for new writers. When they eventually began to appear, reviewers and critics were ready for them, not only welcoming them but defining and confidently generalizing about them. In 1953 John Wain, a 28-year-old poet, critic, and lecturer in English at Reading University, was invited by the BBC Third Programme to put on a monthly literary radio magazine called 'First Reading', with the accent on youth. Wain's concept of youth was sober-sided; he argued that after the modernist excitements of the earlier twentieth century the main literary task was the consolidation of what had already been achieved, without engaging in further experimentation. He was attacked for advancing an unadventurous idea of the possibilities of literature, but it was in tune with a common mood in the national culture, still quietly convalescent after the traumas of war. As Kate Whitehead describes in her valuable book, *The Third Programme*: *A Literary History* (1989), 'First Reading' was widely criticized; an extract from Kingsley Amis's still-unpublished *Lucky Jim*, in which Jim smokes in bed and burns holes in his hostess's sheets, was found particularly objectionable by many listeners. Among the objectors was P. H. Newby, the Controller of the Third Programme, but Wain insisted to him, correctly, that 'it is going to be a famous book some day'.[1]

A harsh review of 'First Reading' in the *New Statesman* provoked an aggressive rejoinder from Wain, and a patient explanatory defence from G. S. Fraser, who found significance in the fact that several of the contributors were 'provincial dons'. The ensuing correspondence touched on such points as whether teachers in provincial universities could be called 'dons', though matters of greater importance were raised: Fraser and John Lehmann argued about the distinction between 'provincialism', which was bad, and the 'provincial tradition', which was good. The provinces were significant in the emerging new writing, both as location and a source of alternative values to those of the capital. Furthermore, academics—whether or not they were called 'dons'—were becoming more involved in both the production and assessment of literature than in the past. University literary study acquired a new prominence, threatening traditional metropolitan *belles-lettres*.

In the autumn of 1953 Wain published his first novel, *Hurry on Down*, to a generally approving reception. Then early in 1954 Amis's *Lucky Jim* appeared and was an instant success, objections to the broadcast episode having been forgotten or overlooked. 1954 was a vintage year for English fiction. It saw the publication of three excellent first novels, which launched their authors in distinguished careers: *Lucky Jim*, Iris Murdoch's *Under the Net*, and William Golding's *Lord of the Flies*. (All three authors eventually received titles: Amis and Golding as knights, Murdoch as a dame). It was also the year of Wyndham Lewis's *Self Condemned*. Reviewing *Lucky Jim*, Walter Allen made a trend-spotting but perceptive attempt to define what was new in Wain and Amis:

A new hero has arisen among us. Is he the intellectual tough, or the tough intellectual? He is consciously, even conscientiously, graceless. His face, when not dead-pan, is set in a snarl of exasperation. He has one skin too few, but his is not the sensitiveness of the young man in earlier twentieth-century fiction: it is to the phoney that his nerve-ends are tremblingly exposed, and at the least suspicion of the phoney he goes tough. He is at odds with his conventional university educa-tion, though he comes generally from a famous university: he has seen through the academic racket as he sees through all the others. A racket is phoneyness organized, and in contact with phoneyness he turns red just as litmus paper does in contact with acid. In life he has been among us for some little time. One may speculate whence he derives. The Services, certainly, helped to make him; but George Orwell, Dr Leavis and the Logical Positivists—or, rather, the atti-tudes these represent—all contributed to his genesis. In fiction I think he first arrived last year, as the central character of Mr John Wain's novel *Hurry on Down*. He turns up again in Mr Amis's *Lucky Jim*.[2]

Later in the year Allen's analysis was quoted and enlarged on in an unsigned article in the *Spectator* called 'In the Movement' (1 October 1954). The author reflected:

The names which meet with approval in cultural circles are still the approved names of the Thirties, and soon they must be quite worn away. They are, and will remain, great names, but as Taste moves on in its clumsy inexorable way the approved names of each generation must necessarily grow dim, and fade, and be four-fifths forgotten, until at last Taste resurrects them in that long run in which we are all dead. Who do you take with you on the long week-ends in Sussex cottages? Kafka and Kierkegaard, Proust and Henry James? Dylan Thomas, *The Confidential Clerk*, *The Age of Anxiety*, and *The Golden Horizon*? You belong to an age that is passing.

The article knowledges that thus far the new movement did not have much to show for it: the first novels of Wain, Amis, and Murdoch, and the poetry of Donald Davie and Thom Gunn (and presumably of Wain and Amis when wearing their poet's hats). But the author is confident that a major shift of taste is taking place. He was not wholly wrong, but the tone of the piece is jarring and it reads like a public-relations handout rather than a serious attempt at literary and cultural analysis.

It was written by J. D. Scott, at that time literary editor of the *Spectator*, and his aim (as he subsequently acknowledged) was to boost the circulation of the magazine by associating it with a new and exciting literary movement; it was already publishing poems and reviews by the new writers. Scott was a good novelist, and the article shows him writing, perhaps with tongue in cheek, well below his proper level. The Movement is said to be:

> bored by the despair of the Forties, not much interested in suffering, and extremely impatient of poetic sensibility, especially poetic sensibility about the writer and society . . . The Movement, as well as being anti-phoney, is anti-wet; sceptical, robust, ironic, prepared to be as comfortable as possible in a wicked, commercial, threatened world which doesn't look, anyway, as if it's going to be changed much by a couple of handfuls of young English writers.

The article succeeded in Scott's aim of arousing interest and was followed by several weeks of correspondence; some of it supportive, rather more of it hostile; his remark about the Movement not being interested in suffering was found disturbing. Such confident generalizations inevitably collapse when pressed, though as someone who was around at the time I have to admit that Scott's formulations did capture something of the contemporary *mentalité*. One result of the article was that the new movement now had a name; thereafter it was always known simply as the Movement, and as such it has passed into literary history. In 1980 Blake Morrison published a scholarly study, *The Movement: English Poetry and Fiction of the 1950s*. It is well informed and well documented and is likely to remain the standard work on the subject. But Morrison's book reflects a common uncertainty about whether the defining criteria of the Movement are intrinsic or extrinsic. That is to say, is a Movement poem one marked by formality, compression of sense, wit, and moral concern, and a Movement novel a story of a young man angrily struggling against a frustrating social environment? If so, one has to acknowledge that these qualities are to be found in many texts outside the milieu and circumstances of the Movement. It may be easier to take extrinsic factors, the situation, social position, and cultural and educational formation of writers, as the defining elements. But in that case one has no way of telling, by critical examination, whether or not a text belongs to the Movement.

One of the most interesting letters provoked by Scott's *Spectator* article came from Fraser. He agreed about the 'new common tone', though he rightly rejected the idea that Murdoch's novel had much in common with those by Wain and Amis. The latter two he said:

both take as their centre of observation an irascible temperament. Moralists have long taught us that anger, like lust, blinds and stupefies. Novelists, however, have long contested this in the case of the concupiscent appetite, but it is a genuine novelty to give the irascible appetite an intellectual respectability. Many writers of an older generation are genuinely shocked by Mr Amis's and Mr Wain's novels because these seem, at least by implication, to justify and even praise a sustained mood of savage exasperation and intolerant bitterness which people of my generation might associate, for instance, with Hitlerism. Yet the irascible appetite has, according to medieval philosophers, its proper object, which is difficulty; and I suppose one might build up a kind of moral justification of Jim Dixon and Charles Lumley by saying that their anger is directed fundamentally against the difficulty of leading a good life in modern society.[3]

Fraser's invocation of medieval philosophers to justify a new literary phenomenon is far-fetched, but the real interest of his letter is that he foregrounds the concept of 'anger', more than eighteen months before the first production of John Osborne's *Look Back in Anger* made it a commonplace. What Fraser says about the novels by Wain and Amis and their heroes applies equally to Osborne's play. If the 'Movement' had been the invention of a highbrow weekly, the cult of the 'Angry Young Men' which took over from 1956 was the product of the popular press, in the wake of the great popularity of Osborne's play. The novels of Wain and Amis were rapidly assimilated to this cult, as were subsequent works of fiction and drama. The phrase was taken from the title of the autobiography of the religious and philosophical writer Leslie Paul, *Angry Young Man* (1951), but that source was soon forgotten. The boosting of the Angry Young Men meant that authors were confused or identified with their characters in popular perception, to their natural annoyance.

Compared with Continental Europeans the British are slow to generate artistic and literary movements. But those that appeared in the past were produced by artists and writers who shared common aims and ideals, which they were eager to propagate, like the Pre-Raphaelite Brotherhood of 1848, the Vorticists of 1914, or the New Apocalypse of 1939. The movements of the 1950s, though they may have grouped together writers who shared a common tone and a common way of looking at the world, were not set up by the collaborating producers themselves, not all of whom knew each other, but defined and labelled from outside by critics, journalists, and publicists.

II

The new novels of the 1950s had some obvious family resemblances, which invited journalistic labelling. They were written in the mode of traditional realism; they presented an irascible hero in a provincial setting, and their perspective and attitudes were lower middle class or occasionally working class. Realistic narration was seen as a reaction against the 'experimentalism', in other words, modernism, which characterized the 1920s, and was notably exemplified by Joyce and Virginia Woolf. The move against 'experiment' was led by C. P. Snow, Pamela Hansford Johnson, and William Cooper, who were friends and literary allies (Johnson was Snow's wife). It has been described in an American study, *The Reaction against Experiment in the English Novel, 1950–1960* (1967) by Rubin Rabinowitz.

The emerging literary culture was to a large extent reactive; novelists wanted to get back behind modernism to Edwardian or Victorian or eighteenth-century models, while poets reacted against modernist *vers libre* or the neo-romantic excesses of the early 1940s, favouring strict forms and a cool, rational tone. These retrogressive attitudes have something in common with the nostalgia for more secure times evident in some older writers, and can, I think, be attributed to the wartime sense of personal and collective disruption, persisting beneath the surface of returning prosperity, and to a pervasive unfocused anxiety about British identity in the postwar world. In some respects the reactive spirit did relate to important shifts in the national culture. The English class-system is notoriously complex, but the central fault-line passes through the middle of the middle class, dividing those who were educated at public schools from those who went to other kinds of school (grammar schools or secondary modern schools in the 1950s, comprehensive schools more recently). The distinction has long been clear, though there is no convenient terminology to describe it; to say that those above the dividing line are 'upper middle and upper class', those below it, 'lower middle and working class', is both cumbersome and inexact, given the many ambiguities in the concept of 'class'. In the following remarks I shall adopt what are, I hope, the simple descriptive terms 'upper level' and 'lower level' to convey this basic educational distinction: 'lower level', of course, covers a wide social spectrum from comfortably-off professional people to the very poor. Early in the twentieth century several major writers were of lower-level origin; in 1910, for instance, Thomas Hardy, Arnold Bennett and H. G. Wells were all active, and the young James Joyce and D. H. Lawrence were emerging. The literary culture of the 1930s and 1940s was, however, to an unusual extent in the hands of upper-level writers, such as the public-school-educated poets and novelists of the Auden generation, and the influential editors of the war years, Connolly and Lehmann. Andrew Sinclair has remarked on the prevalence of Old Etonians at that time; in addition to Connolly and Lehmann, there were Henry Green, Anthony Powell, and George Orwell, Alan Pryce-Jones (the editor of the *Times Literary Supplement*), the thriller-writer Ian Fleming, and the philosopher A. J. Ayer.

The subject-matter of upper-level fiction is likely to be set in a broad but familiar terrain, extending from the acceptable postal-districts of London to Oxford and Cambridge, to the villages and cottages of the Home Counties, to country houses further afield, perhaps as far as Ireland or Scotland, and on to favoured Continental locations such as Paris, Florence, and Venice. English provincial towns rarely come into it, except as occasions for slumming or comic relief. Young men educated at public schools have, as innumerable novels and memoirs have shown, been subject to every kind of humiliation and absurdity, and perhaps suffered psychological damage in the process. But they may also, if they went to the right school with good teachers and stimulating friends, have been well grounded in the arts and high culture generally. They will have been taught languages well and if they have engaged in the traditional upper-level pursuit of foreign travel they will have had the chance to use them. (Though Richard Hillary and Keith Douglas were very young when they died, they had both travelled in Europe in the late 1930s.) Grammar-school boys and girls were less likely to go abroad, for cultural and financial reasons; they may have been quite well trained in languages but not been able to do much with them beyond passing examinations.

The literary movements of the 1950s represented a resurgence of the lower level, with wide cultural implications. Several of the writers involved had, it is true, been to Oxford or Cambridge on scholarships, and thereby joined a meritocratic élite; but entry to an ancient university by no means implies certain entry to the upper level of English society, with its immensely complex web of connections between Oxford and Cambridge, the public schools, country-house culture, the traditional metropolitan institutions of power and authority, and the Church of England. The assertion of provincial virtues was one aspect of the reaction against upper-level metropolitanism. William Cooper's *Scenes from Provincial Life* (1950) looked like a first novel, but in fact its author had published several previous novels under his own name of H. S. Hoff; with this book he launched himself on a fresh career under a pseudonym. Its title makes a double allusion to George Eliot, to *Scenes of Clerical Life* and to the subtitle of *Middlemarch*, 'A Study of Provincial Life'. Cooper's novel is a study of a group of young professional people—a schoolmaster, an accountant, a commercial artist—

and their ambitions and love affairs in a provincial city in the months leading up to the outbreak of the Second World War. It is a charming, observant, and quietly humorous study of ordinary life continuing in the shadow of a major crisis, though it is also a slight, rather inert work. Writers who came after Cooper had a very high opinion of *Scenes from Provincial Life*, for the sense of new possibilities it offered. Malcolm Bradbury has written:

Seminal is not a word I am fond of, wrote one young writer, John Braine, about it. Nevertheless I am forced to use it. This book was for me—and I suspect many others—a seminal influence. And there can be little doubt that not only Braine but a number of other writers, including David Storey, Stanley Middleton, Stan Barstow and indeed myself, were encouraged by it to find a sense of direction in the period after the decline both of modernism and the political fiction of the 1930s.[4]

The unnamed setting of *Scenes from Provincial Life* is, in fact, Leicester, a city which was a recurring locus in the fiction of the period. It was Snow's home-town as well as Cooper's and appears in the early volumes of *Strangers and Brothers*; it also provides the setting of Anthony Burgess's *The Right to an Answer*. A version of its university appears in *Lucky Jim* and in Bradbury's *Eating People is Wrong* (1959).

John Wain's *Hurry on Down* (1953) was the first of the new novels to provide not only a provincial setting but an irascible or proto-angry hero. The central character and consciousness, Charles Lumley, is a young middle-class graduate with a mediocre degree in history, who declines to enter the professional rat-race. He becomes successively a window-cleaner, a driver of export-delivery cars, a drug-smuggler, and at the end of the book he is writing jokes for a radio comedy show. The novel is picaresque, without a clear plot, simply tracing Charles's social and geographical movements and the adventures that befall him. Reviewers compared it to Smollett, and the novel's crude energetic inventiveness gives the comparison some point. Another possible model is Arnold Bennett's comic novel, *The Card*; Wain, a Staffordshire man, admired

Bennett and published a study of him in *Preliminary Essays* (1957). Yet *Hurry on Down* lacks Bennett's geniality. Charles Lumley's bitter mental vituperation against the human obstacles that he encounters recalls Gordon Comstock in Orwell's *Keep the Aspidistra Flying*. The scene in which Charles takes his girl to visit his old college on a fine sunny day is as misanthropic as anything in Wyndham Lewis: 'The undergraduates, who had been chosen from among many applicants on the strength of their intelligence, breeding, and ability to uphold the academic traditions of six centuries, were sitting on the grass in the garden with their shirts off, making clumsy and inexperienced overtures to the giggling maidens who accompanied them.' Charles is not Wain, of course, and authors are not to be confused with their characters, or even with the implied narrator of a novel. Nevertheless, Charles is not presented with any obvious distancing, and his tone is similar to that of Wain's critical and journalistic writing. In some ways *Hurry on Down* is a pre-realistic novel. Wain, unlike Bennett or Orwell. or his contemporaries Angus Wilson, Kingsley Amis, or John Braine, shows little of the realistic writer's interest in cultural appearances and surfaces, the specific look of houses and their interiors, of furniture and artefacts, and what they tell us of the people who live among them. His physical descriptions are as minimal and generalized as those in Fielding.

Amis's *Lucky Jim*, published a few months after Wain's novel, was greeted as the funniest first novel since Evelyn Waugh's *Decline and Fall* (1928), and is now rightly regarded as a comic classic. Jim Dixon is a junior lecturer in history in a provincial university; he does not much like history or academic life, but he needs the job and he is dependent on the favour of his boss, the pompous and evasive Professor Welch, for its continuation. The other major problem is his emotional entanglement with Margaret, a neurotic female colleague. At the end, after many escapades and embarrassments—the crowning one being Jim's drunken public lecture on 'Merrie England'—everything works out well for him. A genial rich man provides him with a job in London, Jim is rid of Margaret and gets the pretty girl he has long fancied, having taken her away from Welch's odious son Bertrand. The basic mode is realistic but the happy ending comes out of romance. The comic element in Amis's writing is not just a matter of describing absurd or farcical events, but is enacted in the writing, as it is in Waugh or P. G. Wodehouse.

Lucky Jim is the work of a subtle literary artist; the prose makes constant play with different registers of language, deliberately using clichés, journalese, academic and bureaucratic jargon, and any other verbal devices that come to hand. Amis is particularly fertile with metaphor and simile. When Professor Welch struggles to grasp a fresh turn in the conversation, his 'clay-like features changed indefinably as his attention, like a squadron of slow old battleships, began wheeling to face this new phenomenon.' This is an expansive simile of the kind found in traditional epic, though relying on the reader's familiarity via films and newsreels with the image it presents. At the same time, it reinforces Welch's general air of archaic ineptness. Later on the same page we read, 'After no more than a minor swerve the misfiring vehicle of his conversation had been hauled back on its usual course.' This sounds a little like Henry James, though in its context it also invokes Welch's old and ill-driven car. Elsewhere Amis seems to be parodying the fanciful similes of Graham Greene's early novels: 'Fury flared up in his mind like forgotten toast under a grill'; 'Hatred lit him up briefly like a neon sign'. Amis's characteristic figures of speech combine wild exaggeration with a sober tone: 'Dixon felt like a man interrupted at his investiture with the Order of Merit to be told that a six-figure cheque from a football pool awaits him in the lobby.' No stylistic devices are by themselves a sign, still less a guarantee, of literary merit, but Amis's concern with language and his continual fine-tuning of verbal effects show his artistic dedication.

Jim Dixon's responses to the world around him are as acute as those of Stephen Dedalus, though he responds not with aesthetic aloofness but with fierce exasperation. Much of the humour arises from the gap between the world Jim lives in and his inner life, which directs a stream of silent mocking commentary at it; the mockery takes physical form in his facial contortions. As David Lodge has pointed out, there is a verbal convergence between inner and outer late in the novel, after Jim's fight with Bertrand: 'The bloody old towser-faced boot-faced totem-pole on a crap reservation, Dixon thought. "You bloody old towser-faced boot-faced totem-pole on a crap reservation," he said.' Amis's comedy arises from a combination of acute observation and absence of empathy. It exemplifies the external approach to humour advocated and practised by Wyndham Lewis, and evident in earlier masters of comedy such as Ben Jonson. There is an element of cruelty in it, which has been discussed by theorists of the comic, notably Bergson; arguably, if we were truly sensitive and humane we would not laugh at people in the way that Jonson or Amis invites us to. For Jim his enemies' personal appearance is a constant provocation, as in Bertrand's asymmetrical beard, or the shape of Welch's nose, 'a large open-pored tetrahedron'.

There are more complex implications in Jim's awareness of Margaret's unattractive elements, such as her lack-lustre hair, the lines on her face, and the thinness of the flesh around her jaw. The presentation of Margaret, who is a neurotic and a liar but a potentially interesting fictional character, is a little troubling. Jim's relationship with her is based on some affection, rather more pity, and a minimal degree of sexual attraction; to have explored it adequately would have involved Amis in a different and darker mode of fiction. Christine, the girl whom Jim gets in the end is no more than a pretty face on a pin-up girl's body. She is a conventional male fantasy-figure, but she also implies an attitude common in the writing of the Fifties (though there are antecedents in Gissing), that stunningly attractive girls are barred to lower-level males, so that Jim's ultimate capture of one is a social as well as a sexual coup. The most sympathetic female character in *Lucky Jim* is Carol Goldsmith, a vivacious older woman who is bored with her husband; she has been having an affair with Bertrand and at the end of the novel seems about to embark on one with Gore-Urquart, Jim's new employer. Amis's understanding presentation of Carol shows that he is not always limited by sexist stereotyping in his creation of women.

In 1955 D. J. Enright published his first novel, *Academic Year*; since that also was an iconoclastic account of university life, it was inevitably compared to Amis's book. But the differences are considerable. *Academic Year* is set in and around the University of Alexandria where Enright had taught in the late 1940s. Two years later Durrell began to put his Technicolor version of Alexandria in circulation with *Justine*; Enright's is recognizable as the same city, but plainer and seedier. The novel has little plot, simply recording the fortunes of three expatriate English teachers over the course of a single academic year. There are a succession of incidents, some farcical, some violent. The comic episodes are very funny, but *Academic Year* is not a wholeheartedly comic novel in the way that *Lucky Jim* is, for the reader is not allowed to forget the grimmer sides of Alexandrian life, with its corruption and poverty. There is a pervading sense that the fairly agreeable and privileged life of the expatriate English is under threat from Egyptian nationalism and will not survive much longer.

Enright is at least as interested in ideas as in events and the characters engage in long discussions, rather like those in the early Aldous Huxley. The novel conveys a similar impression to Enright's poetry and his critical writing, of sardonic but humanly engaged commentary on the casual tragicomedies of existence. It is the work of a highly intelligent writer, who observes closely and feels strongly, but is not a natural novelist. Amis launched the British version of the 'campus novel' with *Lucky Jim*, and Enright gave it an exotic setting in *Academic Year*, but despite the exoticism, he touches on matters that were characteristic of the new British writing. The character obviously closest to the author is Packet, who has gone to university on a scholarship from a poor home, and who reflects on the way in which this experience has alienated him from his family and upbringing: 'the years of unlearning— in a most conscientious and even scholarly way—everything he had learned along with his mother's milk, along with the hissing of the gas mantles, the deliberately muffled tread of boots and slippers on shared staircases, the margarine that was always on the point of being finished up and the crusts of bread that had to be.' Packet is too passive a figure to be called an angry young man, but there are moments when he thinks in the manner of the irascible hero: 'Me—supposed to have been a bright boy, winning scholarships and prizes, accounted a worthy recipient of public monies—and only now, on the way to my thirties, realizing that doors are opened by turning the knob and not by kicking them down—and that if one door is locked there's generally another that's open.' Enright's concern with Packet's origins and background is something that he shares with several of his contemporaries, but which distinguishes him from Amis. In *Lucky Jim* we know little of Dixon's origins, except that he attended a state school, comes from the north-west of England and took a degree at Leicester. In Amis's novel the characters appear as sharply illuminated figures in the foreground, presented with dramatic intensity, whether of action or consciousness, but with little sense of where they come from; this, together with his taste for metaphor, gives Amis unexpected affinities with Henry James.

The next best-selling first novel was John Braine's *Room at the Top* (1957). Like *Lucky Jim*, it was in tune with the feelings of a new generation, but, in contrast, it cannot be called well written. Braine's prose is unadventurously pedestrian except when it rises to the level of advertising copy; indeed, at one point the novel's hero and narrator, Joe Lampton, looking admiringly at a car, reflects, 'It had the tough, functional smartness of the good British sports car; it's a quality which is difficult to convey without using the terms of the advertising copywriter . . .' Joe is working in local government in a northern town just after the war; he has come a long way from his humble origins and is determined to go a long way further, to get to the top literally as well as metaphorically, for all the best people live at the top of a hill overlooking the town. Joe finally makes it by impregnating and then marrying a rich man's daughter, in the process betraying Alice, the older woman with whom he has been having an affair and with whom he found, for a time, real happiness. Braine is at his liveliest when writing about consumer goods, which Joe finds very exciting, and at his embarrassing worst when writing about love or sex.

The story of a young provincial determined to rise in the world is traditional in European literature, particularly in the nineteenth-century French novel. Joe may recall Stendhal's Julien Sorel, particularly in the way he uses women to help his ascent, but the comparison is all to Braine's disadvantage, as Doris Lessing pointed out in an essay published soon after the book appeared:

Stendhal's bitterly opportunist heroes sought their various destinies in the painful twilight of the reaction that followed the French Revolution. The grandeur of Stendhal's vision comes precisely from his bitter knowledge of the pettiness of life after a great vision had failed. But the hero of *Room at the Top*, whose values are similar to Stendhal's heroes, who understand, as clearly as Julien Sorel when he is allowing himself to be corrupted, does not see himself in relation to any larger vision. Therefore he remains petty.[5]

Joe Lampton is 'out for number one', and he presents himself without irony or perspective. His interest as a character is correspondingly limited; even his attempt at remorse at the end, when Alice has killed herself, is too faint to be convincing. Like Balzac and other writers in the great age of realism, Braine is fascinated by property and objects, and as far as possible he provides them with brand-names and price-tags; *Room at the Top* presents a striking early instance of the spirit of postwar consumerism. The association between women and wealth goes back to antiquity, but Joe Lampton updates it in a proto-consumerist epiphany, as he observes a rich young man and his girl getting into the aforementioned sports-car: 'I wanted an Aston-Martin, I wanted a three-guinea linen shirt, I wanted a girl with a Riviera suntan—these were my rights, I felt, a signed and sealed legacy.'

Fired by memories of his family's early poverty, Joe Lampton, like Jim Dixon, is a Labour supporter. So too were their respective creators in the 1950s, but both Amis and

Braine later made the move to the Right that characterized many of their generation. In Braine's case, at least, one could see it signalled in the values that dominate *Room at the Top*. Braine was a naïve writer, without any of the positive qualities, like freshness and unconventionality, that literary naïvety can sometimes bring. In political terms it turned him into what Amis described as 'an extreme right-winger it was embarrassing to be allied with'.[6]

Joe Lampton is not really an Angry Young Man. He has moments of irascibility, usually directed at the people getting in his way, but as he rises in society he learns to control them; one has to avoid annoying those who might become useful. Arthur Seaton, the 21-year-old hero of Alan Sillitoe's *Saturday Night and Sunday Morning* (1958) is frequently angry, and he expresses his anger not in inner tirades, like Jim Dixon, or verbal onslaughts, like Jimmy Porter, but with fists and boots and, on one occasion, with an airgun. This is one of the few novels of the 1950s to have a working-class hero and setting. Arthur works in a Nottingham bicycle factory; he earns good wages, and he spends them on smart clothes, beer, and women. He is a cheerful anarchist, closely attached to his family and mates, and violently hostile to everything outside his immediate circle: employers, businessmen, politicians, bureaucrats, and, significantly, trade-union officials. Arthur is as alienated from organized labour as he is from every other kind of organization, and this sharply distinguishes him from the heroes of the working-class fiction of the 1930s. In Marxist terms, Arthur is a 'lumpenproletarian', devoid of political consciousness, though still voting Labour out of family piety.

In portraying Arthur's alienation from the larger society Sillitoe focuses on an aspect of working-class life that sociologists have analysed and left-wing activists have lamented. But Arthur also seems to have antecedents in the nihilists of nineteenth-century Russian fiction. The descriptions of his work in the factory remind one how rare informed accounts of work, especially industrial work, are in novels, even novels about the working class. In a sense, *Saturday Night and Sunday Morning* is the terminal instance of the nineteenth-century industrial novel, henceforth as obsolete as the kind of manufacturing it describes was shortly to become. But most of the time we see Arthur at play, in the space indicated by the book's title. He comes on at full throttle in the opening chapter, where he falls downstairs in a pub, having consumed seven gins and eleven pints in a boozing contest, which he has won. After he has picked himself up he vomits over two irate customers and rapidly escapes to the arms of his mistress Brenda, the wife of a workmate. Later in the story he gets involved with her sister Winnie, and for a time he is sleeping with both of them. The treatment of sex seems authentic but was startling at the time, particularly Brenda's successful attempt at procuring an abortion with gin and hot baths. At the end, though, Arthur is tamed in time-honoured fashion and is about to marry a nice young girl, Doreen.

The novel precisely catches an historical moment: the early 1950s in England, when industrial production, mainly for export, made a strong though short-lived recovery, and there was full employment and plenty of money about. At the same time, in the novel there is a feeling—proved correct in the long run—that the prosperity cannot last. Memories of the 1930s, when Arthur's father had been through very hard times, contribute a sense of insecurity, added to by worries about the chances of another war. Another significant element noted by Sillitoe is the arrival of television, that harbinger of cultural transformation. Arthur is glad that his father enjoys it but he hates it himself: 'Television, he thought scornfully . . . they'd go barmy if they had them taken away. I'd love it if big Black Marias came down all the streets and men got out with hatchets to go in every house and smash the tellies. Everybody'd go crackers. They wouldn't know what to do.'

Sillitoe presents Arthur as an existentialist hero, asserting the values of selfhood against an inert world. Despite his crudity and violence, his driving energy, so unusual in English writing at that time, comes over as a positive quality. It occasionally threatens to burst the narrative apart, but Sillitoe maintains a firm though unobtrusive control; Arthur is presented with understanding but without collusion. In the pastoral final chapter the tone changes completely, as a chastened Arthur sits quietly fishing and thinking about marriage.

Stan Barstow's *A Kind of Loving* (1960) was another best-selling first novel in the vein of provincial realism, with a working-class setting in a northern industrial town. Sillitoe's characters are confined by their traditional way of life, but Barstow's reflect the social mobility that was becoming common in the late Fifties. The young narrator and protagonist, Vic Brown, is the son of a miner and the family are rooted in working-class customs and values. But Vic has gone to a grammar school and is working as a draughtsman in a local engineering works; his elder sister is a teacher, and his studious younger brother wants to be a doctor. His father is happy that his offspring have gone on to better things. Barstow presents the Brown family and their milieu with warmth and sympathy and a good ear for their ways of speaking. The picture recalls the early Lawrence.

Though Vic likes to sound knowing about sex and has an intense fantasy life, he is still a virgin. He falls violently for a pretty young typist in the office where he works, and she has her own reasons for encouraging him. But his infatuation fades, and when they have been going round together for a while, Vic realizes that Ingrid—her parents named her after Ingrid Bergmann—is superficial and conventional, and that he no longer likes her very much. But by then she is pregnant, after a single fumbled encounter, and according to the still-prevailing mores he has no alternative but to marry her. Barstow conveys poignantly the pain and uncertainty of being young, and the deep mutual incomprehension of an ill-matched couple. The book shows that though the expansive postwar world offered more opportunities to a bright young man like Vic than his father had ever known, the combination of physical drives and restrictive social custom can still blight the fate of the individual, as it had for Hardy's Jude. Vic has to move rapidly from the doubts and hopes of adolescence to a reluctant shouldering of the responsibilities of manhood. Ultimately, the author that Barstow most suggests to me is neither Hardy nor Lawrence but Gissing, in his careful naturalism and his depressed determinism, though with more relieving touches of humour.

Keith Waterhouse's *Billy Liar* (1959) resembles *A Kind of Loving* in a number of ways. Both novels have a youthful narrator (somewhat reminiscent of J. D. Salinger's *The Catcher in the Rye*), and are set in a drab northern town, where young men meet girls in coffee-bars and dance-halls. Billy Fisher comes from a little higher up the social scale than Vic Brown—his father runs a small haulage business—and he seems younger. He works as a clerk at an undertaker's and dreams of being a scriptwriter. The major difference between the two novels is that whereas Barstow's is a domestic tragedy, Waterhouse's is a lively comedy. Billy lives in the fantasy world of a Madame Bovary or a Walter Mitty, and his efforts to get out of trouble by weaving increasingly elaborate lies draw him into ever worse difficulties. *Billy Liar* contains some of the funniest scenes in modern English fiction. Yet the pathos of the provincial youth is also suggested, as in Barstow's book. Vic gets his girl into trouble in the traditional way, whereas Billy is engaged to two girls at the same time and is in love with a third. Both young men are struggling for independence, but live at home and are subject to greater parental control than their counterparts a generation later. *Billy Liar* was turned into a good film—as were the first novels of Braine, Sillitoe, and Barstow—but the film version plays up the farce at the expense of the pathos.

The realism of the 1950s characteristically presented a young lower-level male in a provincial setting, struggling against the barriers set by English class and culture. The struggle was encouraged by the feeling that these barriers were becoming less fixed and absolute. The egalitarianism of the war years was checked in the immediate postwar period, despite the efforts of the Labour Government. But by the mid-1950s increasing prosperity was bringing fresh opportunities for social movement. The new novels reflected changes in society, as fictional realism has always done, though it was excessive to claim (as some did at the time) that the English novel was back on the rails again, having seen off 'experiment'. The realistic novel has always coexisted with other fictional modes, and is likely to go on doing so. This was certainly the case in the 1950s, when mythographers and belated modernists produced important if unfashionable work, which will be discussed in later chapters.

Raymond Williams, in a essay first published in the late Fifties, 'Realism and the Contemporary Novel', found that in many contemporary novels the balance between the personal and the social that characterized the nineteenth-century masters of realism had broken down, and that one had instead 'the fiction of special pleading', in which a personal formula of anger or resentment is projected on to the world. Williams remarks, 'the first-person narrative, on which so much technical brilliance has been lavished, is now ordinarily the mechanism of rationalizing this breakdown', and takes *Room at the Top* as an extreme instance, 'which breaks down altogether, because there is no other reality to refer to'.[7] Williams is right, up to a point; the idea of a formula of special pleading helps to explain the phenomenon of the 'angry' character, or the irascible hero. But writers did not happen to adopt this formula at the same time just because of a coinciding whim. The fact that it became common enough for Williams to identify and analyse says something about the society they were living in and responding to.

III

May 1956 has become a moment of mythic significance in English theatrical history. In that month John Osborne's *Look Back in Anger* was put on at the Royal Court Theatre. The initial critical response was uncertain, recognizing a powerful new talent in Osborne, but unsure what to make of the play. One of the most enthusiastic reviewers, though, was the influential critic Kenneth Tynan, who called it 'the best young play of its decade', presumably referring to the fact that Osborne was at that time only 26, and the four principal characters are in their mid-twenties. It was by no means an avant-garde play like Beckett's *Waiting for Godot*, staged at the Royal Court the previous year. Osborne wrote in a vein

of scrupulous naturalism, with novelistically exact stage-
directions and careful attention to setting and costumes, like
Cliff's new but already badly creased trousers. Although seen
as enacting a revolt against the 'well-made play' that had long
dominated the West End stage, *Look Back in Anger* is itself
a very well-made play with a tight three-act structure, and
appropriate dramatic tensions, expectancies, and resolutions.
Osborne was right when a few years later he described it as 'a
formal, rather old-fashioned play'.[8]

What was new and startling about *Look Back in Anger* was
not its form but its situation and characters, particularly the
central figure of Jimmy Porter, a graduate of a university said
to be not redbrick but white-tile, who has dropped out and
runs a sweet-stall in a Midland town with his Welsh friend,
admirer, and punching-ball, Cliff. The first scene of the play
brought an immediate shock of recognition to contemporary
audiences: Sunday afternoon in a seedy flat, with Jimmy and
Cliff squabbling over the newspapers, and Alison, Jimmy's
upper-class wife, doing the ironing, as she does for most of the
play (prompting one reviewer to remark that she seemed to
have taken in the nation's laundry). The ironing may have
been exaggerated but the boredom of a provincial Sunday was
not. That way of life was familiar to millions but it was new to
the English theatre. Jimmy dominates the play, with his witty,
exasperated tirades against English life and society. The stage-
direction cryptically describes him as 'so vehement as to be
non-committal'. He ruthlessly bullies Alison, whom he nevertheless loves and needs. Arguments about the play focus on
the character of Jimmy: is he an offensive neurotic who needs
professional care, or a rebel without a cause, driven to distraction by the dullness and pointlessness of contemporary
existence? I recall that in the early discussions of the play
there was a division between people who were genuinely
puzzled at what Jimmy was getting so angry about, and those
who thought it was all too obvious; the former tended to be
upper level, the latter, lower level.

Drab and even squalid settings were not unknown on the
stage, but audiences were not used to seeing highly articulate

middle-class people in them. This muddling of class-divisions clearly lay behind some of the hostile criticism. Some critics adopted much the same attitude to Jimmy Porter as Alison's family: her father was puzzled by him and her mother loathed him. Like *Lucky Jim* two years before, *Look Back in Anger* was in tune with new feelings in and about society, which it helped to focus and identify. In Jimmy the uneasy tension between the upper and lower levels, already reflected in novels, turns into open warfare. Alison describes how Jimmy and his friend Hugh had conducted a guerrilla campaign against her family and friends. Jimmy's marriage to Alison raises the topic of 'hypergamy', which was in the air in the 1950s; it refers to a relationship between a man of modest origins and a woman who is his social superior. In Jimmy's campaign against the upper levels Alison is both his ally and his target, and in his verbal onslaughts cultural and sexual politics converge.

Look Back in Anger has endured long after its original occasion—much longer than some sceptics might have expected in 1956—and gets regularly performed and revived. It is theatrically effective and remains interesting to successive generations, despite its faults, such as the sentimental exchanges between Jimmy and Alison and the irresolute conclusion. Later audiences probably take less account of the socio-cultural context, and see the sexual drama in Strindbergian terms, whilst enjoying Jimmy's outbursts as quasi-operatic performances rather than as cries of protest.

Osborne's play was soon assimilated by the media to the novels of Wain and Amis, and the cult of the 'angry young man' was launched. None of these writers expressed themselves in political terms; for Jimmy Porter politics seems to have been something that happened long ago, in the 1930s; culminating in the great cause of the Spanish Civil War, in which his father had fought and been wounded and then came home to die. From the beginning, Osborne's radicalism was ambiguous. The presentation of Alison's father, the Edwardian survivor Colonel Redfern, is remarkably sympathetic, and suggests that Osborne was one of those radicals who, like

Orwell, turn to an ideal version of the recent past to support their attacks on the evils of the present. The note of Edwardian nostalgia is marked in Osborne's next play, *The Entertainer* (1957), which is conceived as a tribute to the old music-hall; in his preliminary note Osborne writes, 'The music hall is dying, and, with it, a significant part of England. Some of the heart of England has gone; something that once belonged to everyone, for this was truly a folk art.' In this play, Osborne intersperses naturalistic drama with the songs and monologues and rapid scene-changes of the music-hall, and to this extent it is formally more varied and adventurous than *Look Back in Anger*. Archie Rice, the battered comedian, a hollow man defeated by life, is the central character, and provided Laurence Olivier with one of his greatest roles. Set against Archie is his dandyish elderly father, Billie, who had been a star of the Edwardian music-hall, and who comes across as far worthier than the corrupt Archie. (Archie is planning to revive the old man's career for his own financial ends, but Billie dies before it can happen.) The trauma of history affects the action, for Archie's soldier son is killed in the Suez adventure. But it is not a visibly political play. In the theatrical devices of *The Entertainer* Osborne may have been influenced by Brecht—whose work had made a considerable impact when the Berliner Ensemble played in London in 1956—but the influence is no more than superficial. Osborne is far from Brecht's stern ideological lesson-drawing.

Other young dramatists were writing realistic plays that explored new forms of life. Colin MacInnes wrote of Osborne, 'by crashing his own way to glory he has left broken gates wide open for so many who came swiftly after.'[9] In the late 1950s and well into the 1960s there was a lively revival of play-writing, centred particularly on the Royal Court Theatre, and the Theatre Royal, Stratford, in East London. The new realism was not universally popular; many theatre-goers still wanted escapism and glamour, and the dismissive term 'kitchen sink drama' was directed at works that seemed too grimly naturalistic. Osborne's ironing-board had been bad enough, but the sink was worse; washing-up is taking place at the opening of Arnold Wesker's *Roots* and goes on for much of the play.

One of the most interesting works to be placed in this category was Shelagh Delaney's *A Taste of Honey*; the author was only 19 when it was first performed at the Theatre Royal in 1958. She had grown up in Salford, failed to get into a grammar school, and left school at 16 to work in a factory. She wrote *A Taste of Honey* after she had seen a Terence Rattigan play on tour and, though she knew nothing about drama, thought she could do better herself. The setting is dismal, a broken-down flat in Manchester, into which move Helen— described in the stage directions as a 'semi-whore'—and her teenage daughter Jo. The action is minimal. Helen moves out to get married; Jo has a brief tender affair, called an engage- ment, with a black sailor, who then disappears. Her mother's marriage does not work and she returns to the flat. Meanwhile Jo is pregnant and is devotedly looked after by a young homosexual friend. Delaney writes excellent dialogue, and Jo establishes herself as a witty and forceful character. Despite the realism and the squalid setting and circumstances, the play imposes an almost fairy-tale quality, with strange, delicately lyrical currents of feeling. The 'real' world, the northern industrial city, England in the 1950s, seem very remote. No doubt Shelagh Delaney was projecting adolescent fantasies, especially in the character of Peter, Helen's attractive but bounderish suitor, but she turned them into art, and her play shows an astonishing natural talent. But it remained a one-off achievement, though a very successful one, with long runs in London and New York, and a popular film version.

The most ambitious of the new dramatists was Arnold Wesker, who for a time achieved a celebrity rivalling Osborne's. Unlike his contemporaries, Wesker was an avowedly political writer, an idealist of the Left with a passionate desire to break down the barriers that separate the mass of the people from the riches of high culture. His three related plays, *Chicken Soup with Barley*, *Roots*, and *I'm Talking About Jerusalem*, were first performed in 1958–60, and published as *The Wesker Trilogy* in 1960. In the trilogy Wesker traces the rise and fall of socialist aspirations, from the anti-fascist commitment of 1936, when the Spanish Civil War breaks out, through the hopes and disappointments aroused by the Labour Government of 1945, the trauma for many Communists of the Soviet suppression of the Hungarian Revolution in 1956, and the reinforcement of Conservative rule in the General Election of 1959. What made Wesker's plays so popular was probably less their political content than their human warmth and vitality, and their understanding of working-class life.

The first and third of the plays deal with the Kahns, an East End Jewish family. When we first meet them and their friends, in October 1936, they are well to the Left, in some cases members of the Communist Party, concerned about Spain, but more immediately about resisting a Mosleyite march into the East End. At this point the Kahns are at the point of their idealistic power; thereafter, we see ideals faltering and commitments changing. *Chicken Soup with Barley*, presents a twenty-year sweep of history, following the fortunes of a group of representative characters, but does so with the techniques of old-fashioned naturalism when it requires something like the model of Brechtian epic drama to be effective. Here, and throughout the trilogy, there is a conflict between Wesker's conscious belief in the importance of politics and his deeper sense that what really matters in human life is the true voice of feeling.

This sense is dominant in *Roots*, which seems to me the best of the three plays, or at least the most dramatically integrated. The Kahns do not appear directly, and the only connection is that the central figure, Beatie Bryant, is Ronnie Kahn's girl-friend. She was working as a waitress in London when she met Ronnie, who was a small boy in the first play, and is now a working-class intellectual with a lot of ideas but not much aim in life. He does not appear, but Beatie is the mouthpiece for his thoughts. She is a lively, hopeful, intelligent girl, whom we meet when she is visiting her family in rural Norfolk. The Bryants·are presented sympathetically, or at least fair-mindedly, but Wesker shows them as extremely stupid and ignorant, illustrating Marx's phrase in the *Communist Manifesto* about the 'idiocy of rural life'. Wesker insists on the Cold Comfort Farm aspects of Beatie's family, perhaps to rebuff any pastoral sentimentalism about the Organic Community, but the dismal picture seems rather exaggerated for the late 1950s. Beatie does her best to introduce them to better things, as Ronnie has her, even trying to get her mother to listen to light classical music on the radio. She does not have much success, and she remains a female version of the figure at odds with her environment who occurs in so much writing of the time. Beatie has something of the compelling energy of Jimmy Porter, though a much nicer nature, and like him engages in operatic outbursts on stage. Feminist critics would probably see Beatie as very much a male creation—in more than one sense—whose ideas and tastes and capacity for feeling have been roused, Pygmalion-like, by Ronnie. At the end of the play, though, Beatie has been cast off by Ronnie, and instead of being shattered she feels she is now a truly independent person: 'I can feel it's happened, I'm beginning, on my own two feet.' This one might regard as at least a proto-feminist conclusion.

Roots expresses Wesker's conviction, in a long line of reforming thought from the Victorian sages onwards, that the masses need to be humanized by exposure to art and high culture generally. This conviction has since been doubly undermined: by the consumerist belief that 'high' culture is just one possible form of culture, and that people have the right to what they want, or think they want; and by the conviction of tougher-minded left-wing thinkers than Wesker that 'high' culture is really about power, and is thus a repressive instrument. Wesker's passionate but naïve idealism is more inviting and responsible than either of these stances. Whether the plays in which he enacted his ideas have the dramatic capacity to renew themselves and stand revival might be worth putting to the test.

IV

Although novels and plays attracted most attention, it was in poetry that the clearly identifiable signs of a new spirit first appeared. As early as 1950 a prescriptive call appeared in *Poetry London*, under new editorial direction after the departure of Tambimuttu (and soon to cease publication, like the other literary magazines which had flourished during the war): 'Poets should take stock of themselves, especially with a view to improving their technical equipment and sharpening their wits. We should like to see poets turn their attention more to satirical verse, or occasional poems with a bite and edge to them . . . What is needed is not so much the "inspired" poem as a revival of *style*: first class workmanship rather than the prophetic tone' (January 1950). And so it came to pass, for within a few years young poets were writing in the way called for by *Poetry London*. Their early work had a limited circulation, as it came out from small publishers, like the press of the School of Art at Reading University, the Fantasy Press (which operated in a village near Oxford), and, a little later, the Marvell Press at Hull. But wider public interest was soon aroused.

In a review-article in the *Spectator* for 27 August 1954, Anthony Hartley attempted to assess the new poetry. It was formal though conversational in tone, cool and dissenting in manner, opposed to Neo-Romantic rhetoric, with a bias towards moral seriousness and intellectual complexity. Poets might treat mythological subjects but were critical of mythopoeic thinking. William Empson was an important influence, and so was eighteenth-century poetry. Hartley cited the titles of recent books of verse as characterizing the poets' attitudes to language and life: *Fighting Terms*, *Mixed Feelings*, *A Form of Words*, *A Frame of Mind*. He remarked, 'Complication of thought, austerity of tone, colloquialism and avoidance of rhetoric—these provide some common ground and some common dangers.' A few weeks after Hartley's article, J. D. Scott's 'In the Movement' cited it as evidence of the new and transforming forces at work in literature. Scott was mainly interested in novels, but the subsequent cult of anger complicated categories. A Movement poem proved easier to identify than a Movement novel, and in later usage there has been a tendency to restrict the term to poetry.

In 1955 the first anthology of Movement-style poetry was published, improbably enough in Japan. This was *Poets of the 1950s: An Anthology of New English Verse*, edited by D. J. Enright, who was at that time teaching in a Japanese university and who compiled it for students of English in Japan. The contributors were Kingsley Amis, Robert Conquest, Donald Davie, Enright himself, Elizabeth Jennings, Philip Larkin, and John Wain. It also included brief statements by the poets about their attitudes to their work, and these provided some vulnerable *obiter dicta*, such as Amis's 'nobody wants any more poems on the grander themes for a few years, but at the same time nobody wants any more poems about philosophers or paintings or novelists or art galleries or mythology or foreign cities or other poems'; or Larkin's 'I believe that every poem must be its own sole freshly-created universe, and therefore have no belief in "tradition" or a common myth-kitty or casual allusions in poems to other poems or poets . . .' The following year Robert Conquest brought out *New Lines: An Anthology* from a London publisher; it contained the same poets as Enright's anthology (and many of the same poems), with the addition of Thom Gunn. Seven of the contributors to *New Lines* were, had been or were to become, university teachers, and the other two— Larkin and Jennings—were librarians. It is not surprising that such a company wrote poetry with an academic flavour, and that the allusiveness and cultural references which Amis and Larkin deplored were widespread. The first poem in the book, Elizabeth Jennings's 'Afternoon in Florence', broke Amis's embargo on poems about foreign cities, while Larkin himself alluded to Tennyson's *The Princess* in his 'Lines on a Young Lady's Photograph Album'. Tradition is inescapable, even if one is fighting against it, and poems are always to some extent about other poems, as Larkin should have realized, and probably did realize in his practice if not in his ideas. Indeed, a reflective self-consciousness about the nature of poetry and the act of writing provided a recurrent note in *New Lines*; Conquest's 'Epistemology of Poetry' is a characteristic instance.

Like other writing of the 1950s, Movement poetry was in reaction against what had gone before, though it was not clear exactly what. Dylan Thomas was the most likely target, though he was seldom mentioned directly; an attitude to poetry that valued craftsmanship highly would have had difficulties with the fact that Thomas was an ingenious and fanatical craftsman, even though his craft was directed to ends the later poets found uncongenial. The New Apocalyptics were the usual whipping-boys, and their importance and influence were greatly exaggerated; indeed, the poetic history of the 1940s has been misrepresented ever since. One would never guess from Movement polemics that poetry showed great variety during the war, when Roy Fuller, Drummond Allison, Norman Cameron, Keith Douglas, and most of the Cairo poets were producing the kind of work approved of by the Movement. In the mid-Fifties, though, the Movement seemed to be doing something genuinely new.

Formally, the Movement represented an attempt to get behind modernism, just as novelists were turning away from 'experiment' and writing in traditional narrative modes. Strict forms were favoured, like *terza rima* and the villanelle, imitated from Empson's use of them, though more commonly a metrically tight, regularly rhyming quatrain, recalling the eighteenth-century hymnologists whom Davie admired. Poems were to be written in the syntax of prose, with a core of rational argument, and metaphor was an optional rather than a necessary component of poetry. The Movement stance was a version of the Apollonian: controlled, rational, sober, understated, opposed to the Dionysiac excesses associated with the Neo-Romanticism of the 1940s (in contrast, too, to the school of anger with which Wain and Amis were identified as novelists). Amis's poem 'Against Romanticism', included in *New Lines*, is a kind of manifesto. Romanticism is temptation to disorder:

> A traveller who walks a temperate zone
> —Woods devoid of beasts, roads that please the foot—
> Finds that its decent surface grows too thin:
> Something unperceived fumbles at his nerves.

> To please an ingrown taste for anarchy
>   Torrid images circle in the wood,
> And sweat for recognition up the road . . .

In its attack on romantic excess the poem implies a secular, agnostic and empirical view of the world:

> Over all, a grand meaning fills the scene,
> And sets the brain raging with prophecy,
>   Raging to discard real time and place,
>   Raging to build a better time and place . . .

The reaction against passion and extremity may seem pusillanimous, and did so at the time to many readers, but those who had been through the war felt that they had seen enough of such excesses. Davie develops the point in 'Rejoinder to a Critic', which attacks feeling because of the crimes it leads to. He opens the final stanza with a line from Donne:

> 'Alas, alas, who's injured by my love?'
> And recent history answers: Half Japan!
> Not love, but hate? Well, both are versions of
> The 'feeling' that you dare me to . . .
> How dare we now be anything but numb?

Davie is rejecting the Romantic endorsement of feeling as self-authenticating good—presented as such in the plays of Osborne and Wesker—in favour of the older 'classical' suspicion of the passions as destructive. Nevertheless, in a later essay Davie acknowledged that Romanticism was an inescapable heritage of the modern writer: 'we are all, like it or not, post-Romantic people . . . the historical developments which we label "Romanticism" were not a series of aberrations which we can and should disown, but rather a sort of landslide which permanently transformed the mental landscape which in the twentieth century we inhabit, however reluctantly.'[10]

The retrogressive posture of the Movement coincided with a critical attempt to rewrite the history of recent English poetry, so as to bracket off the modernism of Eliot and Pound as an alien interlude, and see the true line of development as running from Hardy through Housman, Kipling, Edward Thomas, the best of the Georgians, Wilfred Owen, Robert Graves, Edwin Muir, and John Betjeman. It is significant that in the late 1950s Betjeman became a poetic bestseller. Previously he had been regarded as an odd survival from the 1930s, a true poet, but of considerable affectation and eccentricity, who sedulously imitated minor Victorian poets. But with his new fame the way to a future knighthood and the Laureateship lay plainly ahead. Larkin admired Betjeman, not least for his popular appeal; he also regarded Hardy as the greatest poet of the twentieth century, dismissed modernism as an aberration, and would himself have fitted in happily as the latest poet in a truly English poetic line.

But the Movement poets were not all of one mind on the nature of tradition. Davie, though a *soi disant* 'Pasticheur of late Augustan styles', greatly admired Pound, whom Larkin detested, and did not want to bypass modernism even if it were possible. Davie always had wider horizons than his contemporaries, and though he admired Larkin's poetry found himself at odds with Larkin's insularity and apparent lack of seriousness about poetry.[11] Movement poetry existed as a tenuously coherent entity only from about 1953 to 1956 and was already beginning to fragment when *New Lines* put it on the map. But it produced a number of gifted poets, and that is a sufficient achievement. Even the lesser poetry in the Movement manner has something to be said for it; as Johnson remarked of the Metaphysicals, 'To write on their plan, it was at least necessary to read and think.' Nearly four decades on, the important poets among the contributors to *New Lines* are, I believe, Donald Davie, Thom Gunn, Elizabeth Jennings, and Philip Larkin.

Many readers will share my conviction that Larkin was the finest poet of his generation, and will probably be remembered as one of the best British-born poets of the century. Yet from 1955 onwards there have been dissenters, who resist Larkin's insularity, his traditionalism, and, above all, the way in which his slender *œuvre* is so permeated with negative emotions. In

Lawrentian terms, Larkin is 'anti-life'. The charge against Larkin has been renewed by Margaret Drabble, who regards his poetry—and his novels—as the autobiographical utterances of a very inadequate personality, whose work should be accompanied by a health warning.[12] A pithy rejoinder came from a reader of Drabble's article: 'The bleakest of Larkin's poems cheer me up because they show the language, and a mind behind it, still alive and kicking.' Larkin, like other poets before him, is subject to negative feelings about human life, its limitations and its inevitable end, but transforms them in poetry by the play of mind and language. He is one of those artists who finds beauty in drab material, and, to invoke a line of Empson's, shows us how 'to learn a style from a despair'.

Certainly one should not read Larkin's poetry in the light of his pronouncements and the narrow, philistine persona he liked to project; he was not the first poet to write differently, and better, than his expressed ideas. Larkin abjured modernism, but no one carefully exploring the subtle and often surprising interplay between sense, sound, syntax, and verse form in his poetry can doubt that he had learnt from Eliot and perhaps from other masters, and even picked up something from the French Symbolists whom he professed never to have read. Larkin may have regarded himself as a traditionalist, but if so, he was a sophisticated and deceptive traditionalist. 'Church Going', published in *New Lines*, is deservedly famous, and is, I think, a great poem of universal import, expressing the puzzled reverence of a modern post-Christian on encountering the relics of an age of belief. Despite its large emotions, the poem conveys them by a mass of recognizable detail, as the poet explores the church; not least, his sense of himself as an out-of-place figure:

> Once I am sure there's nothing going on
> I step inside, letting the door thud shut.
> Another church: matting, seats, and stone,
> And little books; sprawlings of flowers, cut
> For Sunday, brownish now; some brass and stuff
> Up at the holy end; the small neat organ;

> And a tense, musty, unignorable silence,
> Brewed God knows how long. Hatless, I take off
> My cycle-clips in awkward reverence . . .

'Church Going' shows Larkin's striking mastery of an elaborate stanza-form, which is a permanent legacy of his early devotion to Yeats, however much he later repudiated Yeatsian rhetoric.

There is a similar mastery in another superb poem from *New Lines*, Thom Gunn's 'On the Move', which is about California motorcycle boys:

> On motorcycles, up the road, they come:
> Small, black, as flies hanging in heat, the Boys,
> Until the distance throws them forth, their hum
> Bulges to thunder held by calf and thigh.
> In goggles, donned impersonality,
> In gleaming jackets trophied with the dust,
> They strap in doubt—by hiding it, robust—
> And almost hear a meaning in their noise.

This stanza shows Movement poetry at its best. Rhyme and metre are strict, and the precise, compact syntax reins in the excitement, just as the leather jackets and goggles of the motorcyclists contain their violent spirits, and the stanza ends with a fine example of what some critics identified as the 'bold Drydenic line'. Gunn is using the boys as a potent symbol, like a Romantic poet contemplating skylarks or swans, seeing them as existentialist heroes, 'The self-defined, astride the created will'. 'On the Move' points to Gunn's own future, for before long he settled in California, where he still lives, and abandoned the Movement manner to become a distinguished practitioner of American developments of poetic modernism.

Davie's, too, has been a career of movement, both physical and intellectual, in contrast to his insular friend and adversary, Larkin. *Contra* Amis, he has written many poems about foreign cities. He was for many years an academic in the United States, first at Stanford then at Vanderbilt, before returning to England on his retirement. Davie did not so much abandon the Movement as grow out of it, though maintaining a devotion to the lesser eighteenth-century poets who were an early model. His later career as both poet and critic has been a continuing close encounter with international modernism, with Pound and Pasternak as points of reference, marked throughout by dedication to poetry as an art, and a continual exploration of styles. His work is not marked by outstanding, five-star poems, and needs to be read extensively in order to get an idea of his quality. *New Lines* contains two of his best early poems, 'The Fountain' and 'Woodpigeons at Raheny', as well as the rather too-often quoted and anthologized 'Remembering the Thirties'. One of the most striking of Davie's early poems is 'The Garden Party', where the socially lower-level poet sadly observes the children of the rich at play. It concludes:

> My father, of a more submissive school,
> Remarks the rich themselves are always sad.
> There is that sort of equalizing rule;
> But theirs is all the youth we might have had.

From the beginning Elizabeth Jennings stood out from the other Movement poets, as a woman in an otherwise male environment—Conquest's introduction to *New Lines* suggested a rather clubman atmosphere—and as a Catholic in a generally secular milieu. She shared with them the prevailing cool, rational tone and liking for formal composition, and she possessed greater technical skill than some of the other poets, particularly in her delicate ear for cadence and rhythm. But her spirit was her own; she was, from the beginning, a poet of observation, perception, and reflection, rather more likely to write about places and works of art and animals than about people. The human dimension appears, when it does, in muted love poems, but solitude has always been her element. Jennings has developed gently but not radically since her early work, and her *Collected Poems*, published in 1986, is a quietly impressive achievement, which offers many variations on the traditional themes of love, memory, time, art, and religion.

V

Walter Allen, in his anatomy of the 'new hero', said some perceptive things about the emergent writing. But he suggests one of the most influential of its sources not in what he says but in his choice of words. Allen remarks of the new hero, 'his face, when not dead-pan, is set in a snarl of exasperation . . . at the least suspicion of the phoney he goes tough . . . he has seen through the academic racket as he sees through all the others . . . a racket is phoneyness organized . . .' Attempting to define a new figure in English culture, Allen makes liberal use of what older readers would have disapprovingly called 'Americanisms': 'tough', 'dead-pan', 'phoney', 'racket'. The young provincial Englishman seems about to be played by Humphrey Bogart. The influence of American popular culture, particularly films, jazz, and crime novels, had been widespread in England since the 1920s and had been the subject of disapproving comment from Leavis and Orwell among others. The influence was reinforced during the war years, by the presence of large numbers of American service-men from 1943 onwards, and more extensively by the radio broadcasts of the American Forces Network which catered for them, and which were widely listened to by young English people. These broadcasts brought American jazz and swing and comedy shows to a wide audience in Britain.

The United States was traditionally regarded by the European Left as the democratic and republican embodiment of the ideals of liberty and equality, and something of this attitude persisted in postwar England, though expressed in cultural rather than political terms. American popular culture, notably movies and jazz, challenged traditional upper-level observances. Older intellectuals went as a matter of course to the theatre and classical concerts, and took their holidays in France and Italy; younger ones were more at home with jazz and the cinema, and dreamed of visiting America. Yet there is still a marked ambivalence about America in the new writers, epitomized in Waterhouse's *Billy Liar*. American speech offers a way of evading the English class-markers, but it remains alien. Billy's friend Arthur has an evening job singing pop songs in a dance hall: 'He always affected an American accent when he sang. I disliked it, but I had to admit it was good.' But later in the evening, 'Arthur's American accent had become so pronounced that it was difficult to understand what he was singing about.' Osborne's Jimmy Porter complains of living in the American century and has nostalgic longings for a vanished England; at the same time, he is passionate about the quintessentially American art-form, jazz. John Wain, in a review-article on Dylan Thomas first published in 1953, falls easily into a kind of American idiom: 'we want a little less gas about Thomas, and some criticism that really talks turkey . . .'[13] Wain took the title of his novel *Hurry On Down* from an American blues song, but in the course of it a negative attitude to American influences appears. Charles Lumley is taken to tea with the family of a working-class girl, Rosa. There occurs an authorial intervention in the spirit of Orwell's cultural criticism, unfavourably comparing Rosa's brother Stan with their father: 'At sixty, Stan would have neither the massive good humour nor the genuine dignity of his father, and already he was immersed in learning the technique of cheap smartness. He talked a different language for one thing; it was demotic English of the mid-twentieth century, rapid, slurred, essentially a city dialect and, in origin, essentially American.' Stan is disapprovingly described as smoking 'a cheap American-style cigarette' whilst he is eating his tea of ham and pickles.

During the war years Hemingway had been a potent American literary influence, most obviously in the tough, laconic sketches of action in the front-line or boring inaction behind it which appeared in magazines. In the 1950s Salinger was another. He is self-consciously invoked in an amusing minor novel, Andrew Sinclair's *My Friend Judas* (1959), set in Cambridge; the undergraduate narrator, Ben Birt, describes his attempt to answer his History Finals:

So I slap down my muck-up in a mess, here a bit of Holden Caulfield, there a bit of Miss Lonelyhearts, all mixed to a mash in my good old British earthenware bowl of a brain-box. But, hell, I've a British lush of a soul too, with Dylan T. tap-tap-tapping his tipples out in my unconscious and mad Manley Hopkins gone, man, gone on the vibes of my imaginings. Scratch a Yankified Limey, however tough his top, and somewhere you'll find a screwball of a Shakespeare. All the muckrakers since Dreiser can't make a sloplover other than a lover of slop.

And all this trouble with words too, Jesus, as if a word weren't just a word, without some shitty semanticist whipping out a swift Wittgenstein to wind you by saying everything you say is bad breath. Define your terms, buster, and even then you'll be blowing just a breeze of hot air. Well, mac, so I do, so I do. I spill out one beautiful balls. The platitude wasn't said that I don't say every day.

The versions of America that Yankified Limeys like Ben Birt constructed for themselves were second-hand, drawn from films, books, and music. As travelling to America became easier many English writers went there on academic assignments or cultural jaunts, and saw the fabled place for themselves. The result was a new fictional genre, the novel about the adventures of the Englishman in America, which became common in the 1960s and 1970s. One of the first literary visitors was Thom Gunn, whose 'On the Move' I have mentioned. Gunn responded more acutely to American experience than most of his contemporaries. In 'Elvis Presley', for instance, he engages in a brief but profound reflection on one of the central icons of American mass culture:

> We keep ourselves in touch with a mere dime:
> Distorting hackneyed words in hackneyed songs
> He turns revolt into a style, prolongs
> The impulse to a habit of the time.

To return to Walter Allen's anatomy of the new hero, and his possible sources: 'The Services, certainly, helped to make him; but George Orwell, Dr Leavis and the Logical Positivists—or rather the attitudes these represent—all contributed to his genesis.' Allen's formulation is worth unpacking. When he wrote, it was less than nine years since the war ended, and several of the new writers—and their fictional characters—had taken part in it. And until the late Fifties conscription for military service was an unwelcome fact of life for young men. Hence Allen's invocation of the Services; the continuous struggle against bullshit and boredom and the arbitrary exercise of military authority had a strong formative effect.

The name of George Orwell has often occurred in this study and it would be hard to overestimate his influence. He was regarded as an embodiment of English decency, bloody-mindedness, and grumbling; a dissenting radical with conservative instincts; an Old Etonian with a working-class suspicion of 'them'. 'Dr Leavis', as he was usually known in his lifetime, was an English cultural nonconformist with different interests from Orwell, but, like him, was at odds with the upper-level establishment. As a practising literary critic Leavis was a close reader who demanded analytical rigour, leaving no room for impressionistic raptures about the beauties of literature. Here, as Allen implies, was one possible source for the new hero's intense suspicion of 'phoneyness'. Leavis's prominence in the 1950s had a broader significance. He was both a literary critic and a university teacher, at Cambridge. The combination is now taken for granted but at that time it was still new. Traditionally, critics had been men of letters, not academics, and their background was probably in Classics or History. Leavis exemplified the critic who taught English Literature in a university as well as writing about it, and who regarded the subject as of central importance for humane values. The quarterly *Scrutiny*, which Leavis edited for many years until it ceased publication in 1953, was very influential in propagating his sense of the importance of literary criticism. 'Eng. Lit.' as an institutional subject was now on the map, in the academy and in the national culture at large.

The 1950s saw the appearance of a new generation of literary critics working in universities; they were far from being Leavisites themselves but they had learnt from Leavis's example, directly or indirectly. They included John Bayley, Donald Davie, Frank Kermode, Richard Hoggart, and Raymond Williams. The last two began their careers in university extramural departments, and were exponents of the kind of literary-cum-cultural criticism that Leavis practised and encouraged in the 1930s, and that originated in the great Victorian debates on what Carlyle called the 'Condition of England Question'. Hoggart's *The Uses of Literacy* (1957) and

Williams's *Culture and Society 1780–1950* (1958) applied the approach of the literary critic to questions about society, and aroused much interest of a cross-disciplinary kind. Hoggart's book examined working-class culture and entertainments in an Orwellian and Leavisite spirit, and deplored (though with an undertow of fascination) the glitter of American-style mass civilization. Hoggart recalled in affectionate detail his own working-class childhood in Leeds, and the book suggests a creative writer *manqué*, close in feeling to the novelists of provincial realism.

In time Hoggart and Williams came in from the often literal cold of extramural teaching and became professors, at Birmingham and Cambridge, respectively. Hoggart later turned to administration, and was for some years an Assistant Director-General of UNESCO. Williams's *Culture and Society* was a more conventionally academic book than Hoggart's, the critical history of a tradition to which it made an important contribution itself. But it launched Williams on a prolific career of writing about culture, society, and politics, in analytical or polemical veins. He also published several novels. Much of his work over the next thirty years, until his death in 1988, involved him in intense and sometimes tormented negotiations with Marxism. Williams is a lesser figure than Sartre or Lukács, but he is the one British intellectual of his generation who invites comparison with them.

The last source of the new sensibility that Allen mentions is 'the Logical Positivists'. Strictly speaking, Logical Positivism was the product of the prewar Vienna Circle, which reduced philosophy either to tautology, as in the propositions of logic and mathematics, or to the clarification and ordering of scientific data. The one outstanding British follower of this school was A. J. Ayer, whose coolly iconoclastic *Language, Truth and Logic* was published in 1936, when he was 26. It came out in a second, revised edition in 1946 and was reprinted, read, and discussed throughout the 1950s. By then the austere rigour of Logical Positivism had been modified, not least by Ayer himself; nevertheless, the term continued to be loosely applied by laymen to linguistic philosophy in general. It is true that in the postwar years English philosophy was predominantly analytical and empirical, concerned with questions of logic and language, and hostile to metaphysics (which Ayer had dismissed as merely meaningless utterance). In this respect it had affinities with the new literary climate, which was sceptical of the rhetorical and the pretentious, unwilling to rise to great heights or sink to great depths of feeling, and resistant to the large questions raised by metaphysics or religion or myth. Realism in literary convention offers a parallel to empiricism in philosophy. Amis's 'Against Romanticism' suggests the atmosphere and tone of the dominant linguistic philosophy, and he has indicated in occasional comments his sympathy with it.

Contraries, though, are always likely to appear, and a surprising bestseller in 1956 was *The Outsider*, a work of speculative thought by a young autodidact, Colin Wilson, who aroused media interest when it emerged that he had slept rough on Hampstead Heath whilst writing in libraries during the day. *The Outsider* is the product of a mind functioning at the opposite pole to the prevailing norms of academic philosophy. It is not a work of analysis or of coherently developed argument; it reads like the notebook of an unusually well-read, serious-minded, and inquiring undergraduate, and in fact Wilson later wrote, 'For years—since I was sixteen—I had kept a journal, running to many volumes, in which my literary opinions were interspersed with notes for stories, observations, etc. *The Outsider* was only a filtering off of these journals.'[14] Wilson's ideas were drawn eclectically from Blake, Kierkegaard, Dostoevsky, and Nietzsche, and added up to a passionate plea for existentialism and non-dogmatic religion, emphasizing the role of the lonely strivers after truth whom he called 'Outsiders'. Despite its lack of coherence and originality *The Outsider* received a warm welcome from influential reviewers, and quickly became a cult book. The moral of this event seems to be that although professional philosophers banished questions about metaphysics and religion, many ordinary readers still wanted to pursue them. Wilson's book caught a prevalent interest in existentialism—it had the same title as the English translation of Camus's *L'Etranger*—which for the most part took only a shallow and debased form; though existentialism did find serious literary expression in the poetry of Gunn and the fiction of Sillitoe, particularly his novella, 'The Loneliness of the Long Distance Runner'. Wilson's *The Outsider* seemed to provide an ontological stiffening for the stance of the angry young men.

Whatever one called them—new heroes, angry young men, outsiders—the fashionable literary characters were all males, and so were their creators. There was, though, the one-off success of Shelagh Delaney in the mode of kitchen-sink drama, and in 1960 a first novel by a woman writer had a comparable success. Lynne Reid Banks's *The L-Shaped Room* had something in common with the work of the provincial realists, though it was set not in a Midland or northern town, but in Fulham, at that time a seedy and unfashionable area of West London. It told a very traditional story, about a respectable girl who gets pregnant out of wedlock, but gave it a contemporary twist. Jane, the heroine, has been an actress and has a good job in public relations. But as she moves into her late twenties she feels increasingly burdened by her virginity; she deliberately and joylessly loses it to an old boy-friend, and pregnancy duly follows. She determines to keep the child, but her rigid widower father turns her out of the family home, and determined to preserve her independence she rents the squalid 'L-shaped room' of the title, somewhere in darkest Fulham. She bravely starts to clean the place up, though she suffers badly from morning sickness, and in time loses her job. The other tenants of the house rally round to help her; there is the homosexual black jazz musician who occupies the next room, the tormented Jewish writer who lives downstairs and who falls in love with her, and a kindly whore in the basement; even the cantankerous and grasping landlady turns out to have a fairly good heart. And Jane and her father are finally and touchingly reconciled.

The L-Shaped Room is a deeply sentimental novel, with affinities with women's magazine fiction, and, in places, with Dickens at his direst. But it had a very popular formula: Jane embodies both traditional female vulnerability and an up-to-date pluckiness and independence. She also offers the piquant spectacle of a middle-class young woman moving into a dismal environment, becoming intimate with socially marginalized people, and receiving a valuable education in the process. A further reason for the novel's great success may have been that Jane's is an early instance of what subsequently became a common situation in society: the young woman who does not cope with her pregnancy either by abortion or an unacceptable marriage, but who deliberately embraces single parenthood. As a fictional treatment of an independent woman's vulnerability, Banks's novel is inferior to novels by Jean Rhys and Rosamund Lehmann from the 1930s, and by Margaret Drabble from the 1960s. The important women novelists who appeared in the 1950s were, I believe, Doris Lessing, whom I have mentioned, and Iris Murdoch and Muriel Spark, who are hardly to be considered realists and who will be discussed later.

Lynne Reid Banks's literary territory might have been described by a sociologist as 'Socio-Cultural Marginalization in Inner West London in the late 1950s'. This, too, was the territory opened up in three remarkable novels by Colin MacInnes, *City of Spades* (1957), *Absolute Beginners* (1959), and *Mr Love and Justice* (1960). They deal with, respectively, black immigrants, working-class teenagers, and whores and ponces. Although they are separate works, without continuity of characters, they are unified by their tone and by MacInnes's fascination with London and the life of its streets, pubs, and drinking-clubs; in 1969 they were reprinted in one volume as *Visions of London*. MacInnes was an Australian homosexual, who thus had a doubly detached though extremely observant view of English life. He was also a splendidly comic writer, whose sympathies were all with underdogs or otherwise marginalized figures; thus in *City of Spades* it is the well-meaning white liberals who are the figures of fun. He was one of the first writers to look at the fast-growing West Indian and African communities in London, which he knew very well. There was a certain idealizing and pastoral element in

MacInnes's presentation of blacks, as there was in his view of teenagers in *Absolute Beginners*. In that novel MacInnes focused on the emergence of the teenager as a new social type, made independent by affluence and the rise of a culture catering especially for teenage desires, particularly in clothes and pop music. The streetwise 18-year-old narrator is a self-employed photographer (with discreet pornography as a valuable element in his trade), who regards his father with affectionate but dismissive patronage. This is in significant contrast to the young provincials of Barstow or Waterhouse who are still torn between imitating their fathers and rebelling against them. MacInnes's hero is not altogether credible, or at least not as a truly representative figure; his monologues, like so much writing of the time, owe a debt to Salinger, and his voice is not always to be distinguished from that of his knowing middle-aged creator. Nevertheless, *Absolute Beginners*, like *City of Spades*, is an entertaining and readable novel of genial high spirits, and very illuminating about its epoch.

I would claim rather more for *Mr Love and Justice*, which seems to me to be a work of literary distinction. It concerns the relationship, at first adversarial but eventually collusive, between a young seaman who drifts into becoming a ponce, and an ambitious young CID man. As always, MacInnes shows himself to be well-informed, this time about the ways of prostitutes and their protectors, and of the police, but the novel is thematically and formally tighter than its predecessors. It combines the patterning of a moral fable with the precise observations of London life that characterize the earlier novels. Admittedly less funny than those, it goes deeper; in its reflections on the nature of justice, and the complex relationship between policing and crime, it has a philosophical dimension. MacInnes's London novels are complemented by his collection of essays, *England, Half English* (1961), which includes both conventional literary studies and explorations in popular culture, with accounts of newspaper cartoons and drinking-clubs and teenage tastes in clothes and music. As an essayist MacInnes is Orwell's natural heir, and in *England,*

Half English he does for the 1950s what Orwell did for the Thirties and Forties.

When MacInnes wrote about the London black community in *City of Spades* it had already been treated in fiction by West Indian immigrant writers. George Lamming's *The Emigrants* (1954) was one of the first; it is of documentary interest, but is not a satisfactory novel, perhaps because he was too close to the experiences he was trying to present. The prose is opaque and laborious, and the construction haphazard. Lamming writes much more rewardingly in his collection of reflective and autobiographical essays, *The Pleasures of Exile* (1960), which is full of insights into the situation of the colonial writer. Particularly valuable is his reading of *The Tempest* as a paradigm of imperialism, made long before this approach became standard in politicized Shakespeare criticism. The outstanding novel about West Indian immigrants, written by one of them, is Samuel Selvon's *The Lonely Londoners* (1956). It provides a fine sense of the quality of life in that community, the constant problems of coping with the white world, and the daily hopes and disappointments of the individuals who have come to London to find a better life. It conveys particularly well the shot-silk mixture of sadness and high comedy in their lives. One of Selvon's particular strengths is linguistic; he moves easily between a modified version of West Indian speech (the original would be unintelligible to outsiders) and forms of Standard English, whether spoken by West Indians or English people. He creates a world through language.

MacInnes, Lamming, and Selvon were all writing about a period when West Indians were regarded as British citizens who could freely settle in Britain. By the early 1960s restrictions on immigration were being imposed. An ugly sign of changing attitudes came in a violent race riot that suddenly erupted in Notting Hill, West London, in September 1958; it provides the climax of MacInnes's *Absolute Beginners* and is discussed by Lamming in *The Pleasures of Exile*.

At the time the 1950s were dubbed 'the decade of anger', but in retrospect the anger looks little more than bad temper and extreme irritability. The violence which unexpectedly irrupted at Notting Hill put it in perspective. In the spring of 1956 Osborne's angry hero made his small-scale, individual attempt to shatter the prevailing calm. In the autumn of that year the forces of history made a violent re-entry, as two major international crises broke out simultaneously. The British and the French, in collusion with the Israelis, invaded Egypt in an attempt to overturn the Egyptian seizure of the Suez Canal. This adventure divided Britain more deeply than any political event since the 1930s, and proved to be an ignominious failure; the Conservative prime minister, Anthony Eden, who had instigated it, was forced to resign. Meanwhile, there had been a national uprising in Hungary against Communist rule, which was bloodily suppressed by Soviet forces.

These events had a transforming effect on attitudes. The failure of the Suez adventure made it clear that British imperialism was finally dead; the British Empire was in any case being rapidly dismantled as former colonies became independent, but many people in Britain had been reluctant to accept the fact. The Soviet invasion of Hungary brought disenchantment to the Left. For a long time, many socialists, whilst disliking much that went on in the Soviet Union, had clung to the idea that it still represented a better society, and provided a beacon of hope to humanity. This illusion had been badly damaged a few months before, when the Soviet leader Khrushchev made dramatic revelations about the repressive nature of Stalin's rule. Nevertheless, when Khrushchev himself acted in a thoroughly Stalinist way by sending the Red Army into Hungary and executing the leaders of the revolution, the illusion was shattered.

The result of 'Suez' and 'Hungary', as they soon became known in popular shorthand, was a return of political consciousness. After Hungary left-wing intellectuals and academics abandoned the Communism to which some of them had been clinging, either as party members or fellow-travellers, and looked to a new ideal of democratic socialism. They started journals such as the *New Reasoner* and the *Universities and Left Review*, which later became the *New Left Review*. There was an attempt to rethink the nature of Marxism, paying attention to the humanistic writings of the young Marx, and to relate Marxism to Hoggart's and Williams's socio-cultural explorations of English traditions. Out of this ferment of ideas the New Left was born, and such high-profile manifestations as the Campaign for Nuclear Disarmament. The Conservatives remained firmly in the saddle, and were returned to office in 1959 under Harold Macmillan with an increased majority (the election results are broadcast in the last act of Wesker's *I'm Talking About Jerusalem*). But the sense that politics was simply irrelevant which had been general in the early Fifties had gone; for the first time since 1945 the Left were getting their ideas together.

7

The Myth Kitty

The major modernists had been much concerned with myth: Joyce drew on Homer, Pound on Ovid, and Eliot found a strange excitement in Sir James Frazer's vast compendium of primitive myths, *The Golden Bough*. *Ulysses*, *The Cantos*, *The Waste Land* all reflect these interests. So when Philip Larkin dismissed any belief in a 'common myth kitty' and Kingsley Amis placed mythology alongside foreign cities among the subjects that he hoped nobody wanted any more poems about, they were, in the spirit of the age, implicitly rejecting a central element in the modernist enterprise. Nevertheless, in the early postwar years two extraordinary works appeared from writers marginal to the main tendencies of literary development in England: Malcolm Lowry's *Under the Volcano* (1947) and David Jones's *The Anathemata* (1952), both of them inspired by the high aims of modernism and permeated with myth.

Lowry was born in 1909 in a well-off Cheshire family, and went to Cambridge when the university was a place of particular literary and intellectual fertility, where he got to know I. A. Richards and William Empson. But most of his later life was spent out of England, in Mexico, the United States, and Canada; for several years he lived in poverty in a waterside shack in British Columbia while he worked on the many drafts of *Under the Volcano*. Though a work of less than 400 pages, this novel is so dense and multilayered that one can give little sense of it in a brief account. It is closer to *Ulysses* than any other novel by a twentieth-century British writer, though Walter Allen has pointed out affinities with the work of Joyce's American admirer, Faulkner. Like *Ulysses*, it deals with a few characters in one place during the course of a single day. The place is a remote town in Mexico, the time is the Day of the Dead in November 1938; the principal characters are Geoffrey Firmin, the former British consul and an alcoholic, his ex-wife Yvonne, and his half-brother Hugh. It proves to be the last day of Firmin's life. *Under the Volcano* resembles *Ulysses* in having a firmly naturalistic dimension. Geoffrey Firmin stands out as a substantial and memorable character, as does Leopold Bloom: his drinking—based on Lowry's own experience—is heroic and destructive and produces some wild comedy; like *Ulysses*, much of the novel is very funny.

In its realistic aspect *Under the Volcano* is engaged with the politics and history of its time; there are frequent references to the Spanish Civil War, then in its final phase, which Hugh has covered as a journalist. One year later, when the novel opens in a flash-forward from the main events, the Second World War has already started. But realism, characterization, history, provide only one aspect of the novel. They are balanced by its literary convolution and encyclopaedism, Lowry's tireless pursuit of symbols and allusions, parallels and inter-textual implications; this, too, is Joycean, as was his total dedication to his art and his willingness to undergo personal hardship whilst perfecting it. Underlying the intricacy of detail, certain larger mythic patterns recur, in references to Dante's Inferno and Marlowe's Dr Faustus. The work is a study of hell, set in a landscape overshadowed by two volcanoes, and at the deepest level of its structure it invokes the ancient myth of Tartarus, reputedly located under Mount Etna.

Whatever Lowry learnt from Joyce, and from earlier masters such as Melville and Conrad, *Under the Volcano* remains a work of high originality as well as great artistic power. Allen has said that 'Under the Volcano stands uniquely by itself as a great tragic novel, a masterpiece of organization and of elaborate symbolism that is never forced or strained but right, springing largely as it does out of the scenes in which the action takes place.'[1] I am not quite as sure as Allen that all the formal elaboration of the novel works and is justifiable; some of it may be excessive. Since it was first published, *Under the Volcano*, like other major modernist texts, has become a favourite object of academic explication, and this approach tends to value complexity for its own sake. Nevertheless, when set beside the huge achievement of the novel, these doubts seem unworthy. *Under the Volcano* has earned the right to be called 'great', just as *Ulysses* has, and for similar reasons. Its relationship to the British literature and culture of the time remains problematical; Lowry was born in England and died there in 1957, but he did not share in the experiences that affected so many of his generation. The Canadians have quite a good case in claiming Lowry as a Canadian author and *Under the Volcano* as a masterpiece of their national literature; Lowry's later and much lesser novel, *October Ferry to Gabriola*, published posthumously in 1970 in an unrevised version, is set in Canada.

David Jones was a Londoner of modest Welsh origins whose only higher education was at an art school, and who as a very young man served in the First World War. His experience of the trenches profoundly affected him, and in many respects remained his only really significant physical experience; the rest of his life was quiet and reclusive, marked by recurring ill health. But from the 1920s onwards he established a considerable reputation as a painter in water-colours and as a designer and engraver. Meanwhile, he was engaged on an extended literary work in both verse and prose which arose from his wartime experience, but extended far beyond it. This was published in 1937 as *In Parenthesis*, and is one of the finest imaginative works to have arisen from the First World War, though Jones was right to regard it as much more than a 'war book'. It is also one of the few outstanding modernist texts by a writer born in England. Jones was influenced by Joyce and Eliot—and perhaps by the Cubist painters—in fragmenting language and perception, placing the fragments in unexpected juxtapositions, in a form of literary collage. Above all, like the other modernist masters, he interwove and contrasted past and present, setting the experiences of the front line of 1916 against the battles of the so-called Dark Ages in Britain, a period to which his imagination continually returned.

Jones was a Catholic and very conscious of his Welsh inheritance. He found these two strands converging in the Romano-British civilization that persisted for a time after the departure of Roman power. This culture appealed to Jones by being Celtic, Catholic, and Latinate; it gave birth to the potent myth of King Arthur, the defender of the Romano-Britons against the Saxon invader, and Jones's writings provide some of the latest and most complex manifestations of the Arthurian strain in our literature. He was fascinated by myth, though by native rather than classical mythology. Jones was an autodidact, widely though patchily read in literature, history, mythology, theology, and archeology, and in his consciousness these subjects were interpreted in terms of, or used to interpret, his wartime memories, which he turned into a personal mythology. (In the broadest sense, 'myth' can be defined as the stories which humanity as a whole, or cultures, or individuals, tell in order to make sense of their experiences.) Jones was heavily influenced by the Neo-Thomist aesthetics current in the Eric Gill circle, which he was part of for several years. This treated art as formal rather than expressive, the achievement of man as a maker of artifacts and social rituals and, above all, of symbols. This approach had something in common with the modernist stress on 'impersonality', and Jones's debt to both intellectual traditions made him reject the Romantic belief in self-expression. Certainly, his major texts, *In Parenthesis* and *The Anathemata* (1952), both aspire to a condition of epic impersonality. Yet if *In Parenthesis* is the superior work, as is generally assumed, that may be because it is unified by the thread of Jones's personal experience, embodied in his persona, 'Private Ball'.

The challenge of *The Anathemata* begins with its strange title. Jones tells us that it is to be pronounced with the stress on the third syllable. 'Anathema' usually means a curse or proscription, but Jones is reviving, in the plural form, an older sense of the word, meaning things that are specially preserved, treasured, and offered up. The work is, in effect, Jones's attempt to collect, out of his memories and reading, the treasured fragments of what he thought of as valuable in western civilization. The subtitle is equally challenging and

significant: 'Fragments of an Attempted Writing'. It recalls
what is perhaps the most crucial line of Eliot's *The Waste
Land*: 'These fragments I have shored against my ruins', which
can equally apply to *The Anathemata*. But it also echoes
the subtitle of Eliot's *Sweeney Agonistes*: 'Fragments of an
Aristophanic Melodrama'. The cultivation of the fragment as a
literary mode was a Romantic practice that persisted into
the heart of modernism. And when Jones calls his book an
'attempted writing' he is deliberately evading the question of
whether it is poetry or prose. In fact, modernism had undermined the distinction: Joyce is much more thoroughly a poet
in *Ulysses* and *Finnegans Wake* than in his minor verse com-
positions. In this sense Jones was certainly a poet, and though
The Anathemata (like *In Parenthesis*) is written in a combination of verse and prose, it makes sense to think of it as a long
poem. Nevertheless, even in the undifferentiated poetry of
modernism there are distinctions between the writing of those
on the verse side of the erstwhile divide, like Pound and Eliot,
and those on the prose side, such as Joyce and, I believe,
Jones.

The epigraph to *The Anathemata* is spoken by the Fool
in *King Lear*—'This prophecy Merlin shall make for I live
before his time'—and the words look back to the Arthurian
world that extends from authentic history and archaeology to
mythology. Their context invokes the wholly mythical realm of
the legendary British kings, and the words' vertiginous sense
of moving back and forth in time is altogether appropriate
to Jones's enterprise. The closest literary parallel to *The
Anathemata* is in Pound's *Cantos*, though Jones had not read
them when he wrote it. Like Pound, he voyages through time,
space, and cultures; less widely in space, for he keeps to
Europe and the Mediterranean world, but much more remotely
in time. In the opening section, 'Rite and Fore-Time', he
considers the practices of prehistoric man and the way in
which they prefigure later cultural forms; and then looks
further back still, to the geological records of the forming
planet. But the imaginative heartland of Jones's 'writing'
is Christian Romano-Britain, where memories of Empire

persisted, to be revived in the Middle Ages. Later sections emphasise the fact that Britain is an island, once circum-navigated by Pythias, trading with the rest of the world from the London Docks, near which Jones grew up, through the 'Middle Sea'—the Mediterranean—and the 'Lear Sea'—the English Channel. Throughout *The Anathemata* there are recurring references to the death and resurrection of Christ. Though some Christians had been upset by the discoveries of anthropologists that earlier analogues of those events were to be found in other cultures, Jones was not one of them. He combined religious faith with an evolutionary world-view, and was happy to accept such analogues as foreshadowings of the reality and mystery of Christianity. The Mass is the central event in Jones's conception, anticipated in pre-Christian ceremonies, and in its essence re-enacting the Easter events. The concluding lines of the work describe the priest saying Mass:

> He does what is done in many places
> what he does other
>   he does after the mode
> of what has always been done.
> What did he do other
>   recumbent at the garnished supper?
> What did he do yet other
>   riding the Axile Tree?

The Anathemata is impressively bold in its imaginative sweeping through time; some passages are extraordinarily powerful, and others are quietly moving. But it is, overall, a very difficult work to assimilate. Like other modernist writers, Jones proceeds by a mixture of fragmentation and association. This worked well in *In Parenthesis*, where, as I have suggested, the fragments were brought into coherent relation by the thread of personal narrative. But with *The Anathemata* there is no obvious centre, and the associative links between fragments—as Jones concedes in his notes—can be private and arbitrary. The reader often moves into a dense cloud of references and observations which are evidently important to

Jones but which are not given the formal transformation that he believed to be essential to art. Admittedly he provides notes, unlike Pound, but though they explain references they cannot constitute an artistic justification. And in the passages that work the effect tends to come from what is being conveyed, in striking insights and illuminations, rather than in the way it is conveyed. This is what I had in mind in suggesting that though Jones is the kind of imaginative writer who can properly be regarded as a poet, his poetry is in prose rather than in verse. Much of *The Anathemata* is set out in free verse, but this seems to be a matter of notation, of isolating and presenting the right emphases, rather than of prosodic necessity. Jones, in short, did not have a poetic ear. There is an immediate contrast here with Eliot, whose lines etch or sing themselves into the memory, and, more immediately, with Pound. *The Cantos* are as rebarbative as *The Anathemata* and contain even more of a collage of disparate fragments; but the whole work is permeated by Pound's fine sense of rhythm, heard in the wave-like dactyllic movement that runs throughout; even, somehow, in chunks of quoted documents. There is no comparable unifying rhythm in *The Anathemata* to balance its centrifugal tendency.

If *The Anathemata* is a failure—and I think that, compared with *In Parenthesis*, it has to be regarded as one—then it is the kind of heroic and monumental failure that should be acknowledged and not ignored. Jones was admired in his lifetime by Eliot, who published him with Faber, and by W. H. Auden; after his death his Catholic friends Harman Grisewood and René Hague proved themselves devoted editors and exegetes of his work (Hague published a detailed commentary on *The Anathemata* in 1977). But in general little notice has been taken of his ambitious 'attempted writing', either when it appeared—admittedly a strange literary phenomenon for England in the early 1950s—or subsequently. *The Anathemata* deserves more and better attention, whatever one's ultimate judgement on it. If Jones had been an American it would probably have received it, for modern American poetry is dominated by long poems, by Pound,

Williams, Olsen, Zukovsky, among others, which are all taken
seriously.

David Wright has placed *The Anathemata* in a group of long
poems which he calls 'the major poetic efforts of our century',
though he is careful to distinguish 'effort' from 'success'. They
include Eliot's *Waste Land*, Pound's *Cantos*, Williams's
Patterson, and Hugh MacDiarmid's *In Memoriam James
Joyce*. They are all, he says, concerned with civilization, and
employ the modernist technique of collage to incorporate
passages from other texts.[2] MacDiarmid and Jones are the
only British-born poets in the group, and they present interesting similarities and contrasts, a Scottish Communist and
a Welsh Catholic, both admirers of Joyce, and both with
encyclopaedic interests. MacDiarmid had the poetic ear that
Jones lacked, but *In Memoriam James Joyce* (1955) makes
more tedious reading than *The Anathemata*, being a monologue with little stylistic variety. Despite the nominal allegiance
to Joyce, MacDiarmid is mainly concerned with himself and
his ideas, as he engages in extended reflections and dis-
quisitions on language, philosophy, physics, orientalism, and
other forms of learning. It recalls the flatter stretches of both
The Cantos and *The Prelude*, though MacDiarmid exhibits an
egotistical irritability rather than the Wordsworthian egotistical
sublime. There are, however, moments of poetic intensity that
suggest the fine poet MacDiarmid was at his best, as in his
Scots lyrics and his earlier and more successful attempts at the
modernist long poem, *A Drunk Man Looks at the Thistle*
(1926) and 'On a Raised Beach' (1934).

Two poets who made conspicuous use of myth, and themselves generated a personal mythology, were Robert Graves
and Edwin Muir. They were born at the end of the nineteenth
century, and were traditionalist in their poetic practice though
aware of modernist innovations—Graves unsympathetically,
Muir sympathetically. They were sometimes named in the
1950s and 1960s as representatives of the non-modernist strand
in modern British poetry running from Hardy to Betjeman
and Larkin. Graves had fought in the First World War and
had known Siegfried Sassoon and Wilfred Owen; indeed, he

had been one of the 'trench poets' himself, though he later suppressed his poems about the war, and his major contribution to its literature is his classic autobiography, *Goodbye to All That* (1929). Graves was essentially a love poet, and a very fine one, though for him addressing a poem to a woman was at the same time an invocation of the female principle that was at the heart of poetic creation. Like Hardy, he went on copiously writing poetry well into old age; it was notable for craftsmanship and music, wit and tenderness, but the later poems tend to be variations on limited themes and are not easily distinguished one from another. In his dedicated pursuit of poetry as a sacred calling—though he wrote novels and other prose books to support his life as a poet—Graves was a late inheritor of a central Romantic stance. His poems were usually brief, well-wrought lyrics where intensity of feeling is contained by a cool, restrained tone. This quality made him congenial to the poets of the Movement, and he influenced their work. The terse moral and verbal clarity of Graves's well-known 'In Broken Images' strikes a note that would have been admired and imitated.

Graves's interest in myth, however, was remote from the new mood of the Fifties. Having long pursued it in poetry, in 1948 he published a substantial prose work, *The White Goddess: A Historical Grammar of Poetic Myth*. It is a densely learned but eccentric study which examines myths and folklore and poems from many peoples and periods—paying special attention to the early Welsh poetry that David Jones also drew on—in order to illustrate Graves's conviction that true poets are worshippers of the Moon Goddess, the archetypal embodiment of the feminine. He thereby mythologizes the poetic vocation itself. The book's dedicatory poem to the Goddess opens:

> All saints revile her, and all sober men
> Ruled by the God Apollo's golden mean—
> In scorn of which I sailed to find her
> In distant regions likeliest to hold her
> Whom I desired above all things to know,
> Sister of the mirage and the echo.

Muir was a quiet, reflective, visionary poet. He was also a distinguished critic, and with his wife Willa he made the first translations of Kafka into English. Although he was writing poetry from the 1920s onwards he did not become well known as a poet until he was in his sixties; the titles of his last two collections, *The Labyrinth* (1949) and *One Foot in Eden* (1956), illustrate his mythic preoccupations. Muir traced a myth in his own life. He had enjoyed an idyllic childhood in the pre-industrial island setting of Orkney, which was traumatically shattered when the family moved to the urban hell of late Victorian Glasgow. He found this experience a loss of Eden, as he describes in his beautifully written autobiography, *The Story and the Fable* (1940), later enlarged and reissued as *Autobiography* (1954). The loss always haunted him; sometimes he saw it, as Wordsworth might have done, as an inevitable end of childhood innocence, but he explored it most fully in religious terms towards the end of his life in *One Foot in Eden*.

As a poet Muir was at variance with some of the common assumptions of the past two centuries, notably, that poetry should focus on the specific and particular. His imagination dealt in generalities and was stirred by symbols and archetypes and Platonic ideas, so that the typical title of a Muir poem is a noun, whether abstract or concrete, preceded by the definite article; 'The Way', 'The House', 'The Return', 'The Road', to name only a few. The poems are always carefully written, strongly felt, and not infrequently moving. But their language is rarely memorable, and the remoteness of Muir's poetic mode presents an obstacle to many readers. Encountering symbols of the Incarnation in Italy led him to Christianity, but his poetry lacks an incarnational quality, compared to that of Hopkins or even of the later Eliot; *Four Quartets* is religious poetry with a high degree of abstract reflection, but the language and imagery keep closer to the phenomenal world than Muir was able to. There is something paradoxical about the notion of Platonic poetry, which is essentially what Muir wrote. Nevertheless, in some poems his presentation of the mythic and the archetypal strikes with strange and disturbing force, as in 'The Horses' and 'The Combat'. The latter describes a fight between a fierce heraldic creature, 'Body of leopard, eagle's head | And whetted beak, and lion's mane' and 'A soft round beast as soft as clay; | All rent and patched his wretched skin'. The former attacks the latter but can never quite win. The poem ends:

> And all began. The stealthy paw
> Slashed out and in. Could nothing save
> These rags and tatters from the claw?
> Nothing. And yet I never saw
> A beast so helpless and so brave.
>
> And now, while the trees stand watching, still
> The unequal battle rages there.
> The killing beast that cannot kill
> Swells and swells in his fury till
> You'd almost think it was despair.

The encounter has symbolic implications too multiple to be easily defined and delimited.

One new writer of the 1950s was strongly committed to the fiction of myth and fable. In 1954, a few months after *Lucky Jim* appeared, William Golding published his first novel, *The Lord of the Flies*, at the age of 43. It is a work in which myth is central. As is well known, it describes the fortunes of a group of English schoolboys who are marooned on a desert island. The governing myth of the book is not from classical mythology, but is what has been called 'the English island myth', of which earlier manifestations were *The Tempest*, *Robinson Crusoe*, and *Treasure Island*. Golding was prompted to write by a once-famous Victorian expression of it, R. M. Ballantyne's *The Coral Island*, which shows a group of similarly marooned boys working together to produce a convincing simulacrum of imperial civilization. Golding, taking a conservative and pessimistic view of the possibilities of human nature, shows the boys relapsing into cruel barbarism. In doing so he is not merely expressing a personal stance, though he knew a lot about boys from having worked for many years as a school-master, but is reflecting the shocked sense of the human capacity for evil that became common after the Second World War, in the wake of Auschwitz and Hiroshima. Indeed, *Lord of the Flies* is set against the background of a third world war. Yet Golding's myth also has a traditional religious dimension, as the book's title shows; in the Bible the 'lord of the flies' was Beelzebub, one of the names of the Devil.

The Lord of the Flies became world-famous, and with good reason. It provided a gripping and exciting narrative, which made it popular as a children's book, as *Animal Farm* had been, whilst at the same time raising profound questions about the nature of civil society and the limits and possibilities of human nature. (Its pessimism, though, made it unpopular with progressive humanists.) Nevertheless, *The Lord of the Flies* has the limitations of a moral fable, with elements of both rigidity and thinness; Golding seems to have had a very clear idea in his own mind of what everything stood for, and the underlying ideas are too explicit. His second novel, *The Inheritors* (1955), seems to me distinctly superior. Here, too, Golding is making an intertextual encounter with a literary embodiment of Victorian progressivism, in this case H. G. Wells's 'A Story of the Stone Age', which showed the supersession of Neanderthal Man by Homo Sapiens as a necessary and triumphant step in the ascent of humanity. Golding describes this situation from the point of view of the Neanderthalers, and writes, with immense subtlety and delicacy, from within the limited consciousness of one of them. We see a people lacking in ratiocinative power, but finely intuitive, in touch with each other's minds and with the rest of nature. They are a gentle race, and facing the New Men, who have weapons and developed intelligences, they are doomed. In one aspect, *The Inheritors* can be read as an allegory of colonialism, where the primitive aggressors and expropriators are not Europeans but Homo Sapiens as such. Yet Golding is also making an original conflation of two complimentary myths: the loss of innocence in Eden, and the evolutionary rise of the human race (a powerful myth as well as a matter of scientific fact). *The Inheritors* is written with all the narrative skill that made *Lord of the Flies* so successful, but is more suggestive and poetic in its writing, and more profound in its implications. Golding continued to write, steadily but not prolifically, in a broadly fabular mode, and in 1983 his career was crowned with the Nobel Prize for Literature.

8

Contrary Voices

I

During the 1950s critics and journalists publicized the attitudes associated with the Movement and the Angry Young
Men, which were presented as realist, formally conservative,
empirical, irascible, nonconformist (though secular), insular,
provincial, socially lower-level. Yet some of the most interesting, significant, and generally admired work by new writers
was contrary to these norms. William Golding's mythic and
implicitly religious *Lord of the Flies* was one instance, and
so was another first novel published in the same year, Iris
Murdoch's *Under the Net* (1954).

Because it had a picaresque, *declassé* hero who travelled
light through the contemporary scene, *Under the Net* was
at first regarded as a further instance of the kind of novel
recently published by John Wain and Kingsley Amis, as J. D.
Scott did in 'In the Movement'. It soon became clear, though,
that *Under the Net* was very different from the emerging fiction
of the Movement. Its hero, Jake Donahue, was certainly not
irascible; rather, he was cool to the point of seeming affectless.
He is a bohemian intellectual who suffers from 'shattered
nerves', he translates French novels for a living, and is something of a philosopher. Iris Murdoch at that time taught
philosophy at Oxford, and her first book was a study of Sartre.
So it is not surprising that there is a preoccupation with
philosophy in *Under the Net*; Jake is a kind of existentialist,
with a hint of Colin Wilson's Outsider, though blessed with a
sense of humour, while his friend Dave is a rigorous linguistic
philosopher. Another major character, the mysterious Hugo
Belfounder, a businessman who manufactures fireworks and
produces films, seems to be based on Wittgenstein. Malcolm

Bradbury once argued that the 'net' of the title refers to Vulcan's net in the story of Mars, Venus, and Vulcan, giving the novel a mythological underpinning; Murdoch, however, denied any such intention, saying that she was alluding to what Wittgenstein in the *Tractatus* called the 'net of language'.[1] However, she makes a specific allusion to this myth in a later novel, *A Severed Head* (1961).

In some ways *Under the Net* is a realistic novel. There is the careful presentation of London settings, especially the western postal districts of which Jake remarks, 'There are some parts of London which are necessary and others which are contingent. Everywhere west of Earls Court is contingent except for a few places along the river. I hate contingency. I want everything in my life to have a sufficient reason.' Yet despite the concern with topography and the almost obsessive listing of street names, *Under the Net* conveys little feeling of the sense of life in London, especially if one compares it with Colin MacInnes's novels. In *Under the Net*, and throughout Iris Murdoch's later work, reality seems a matter of invention and projection rather than of felt experience; a 'virtual reality', in the language of recent computer technology. The thin realistic surface of *Under the Net* encloses a sequence of bizarre, comic, or fantastic incidents, presented in a spirit of brisk improvisation. There is the abduction of a film-star dog in its cage from a London flat, a swim in the Thames at dawn, a political demonstration in a film studio which turns into a riot, all incidents presented in great physical detail, as though to enforce the reader's belief in what basically resists it. One of the most memorable and poetically evocative episodes is Jake's pursuit of an old girl-friend across Paris on the evening of the Quatorze Juillet, with a dazzling firework display in the sky. Both in its prose and its setting this scene shows Iris Murdoch's distance from the other emerging novelists of the time; romantic Paris was always a favourite location for upper-level writers. The unexpected combination of realism and fantasy and extravagant inventiveness, and the gentle wit of Jake's observations about the world, made *Under the Net* an instant success, and launched Iris Murdoch on her career as one of the leading authors of her generation. Her innumerable later novels are more complex, and also more generally upper-level in ambience and attitudes than *Under the Net*, but they offer a basically similar blend of apparent realism and actual fantasy.

Muriel Spark also juxtaposes the real and the fantastic. She is of much the same age as Iris Murdoch, and since she published her first novel, *The Comforters*, in 1957, their careers have run approximately parallel, though Spark is less prolific. She had already made a name as a critic and biographer, and her first novel describes a similar London milieu to Murdoch's. If *Under the Net* reflects its author's interests as a philosopher, *The Comforters* is pervaded by Muriel Spark's literary-critical preoccupation with the nature of fiction-making. Its heroine, Caroline Rose, is herself a critic and is writing a study of form in the English novel; at one point she significantly remarks that she is having trouble with the chapter on realism. Caroline starts having aural hallucinations, in which she hears the noise of a typewriter and a voice describing her and what she is doing; these comments are in fact the immediately preceding words of Spark's text. Caroline, it appears, is both a 'real' person and a character in a novel which someone is writing: as indeed she is, and the novel is *The Comforters*. Spark's novel is a brilliant early instance of the writing that displays and explores its own fictionality, subsequently categorized as 'metafiction'. A generation later it has become a commonplace, both as a critical concept and as a fictional mode, but in the Fifties there were few accessible antecedents, apart from Sterne's *Tristram Shandy* and André Gide's *Les Faux Monnayeurs*.

Like Muriel Spark herself, Caroline is a recent convert to Catholicism, and this provides a central element in the story. Spark is a witty and observant commentator on social mores, particularly the behaviour and attitudes of English Catholics in those years. The link between religion and metafiction is that the novelist's relation to the story is analogous to God's to his creation (a word freely used in both religious and artistic contexts). As Ruth Whittaker puts it, 'God, like the novelist, knows the beginning and the end, and the struggles of his characters to escape their destinies, that is, the process of most people's lives, are watched by Mrs Spark with cool, ironic amusement.'[2] Muriel Spark is sufficiently a realist in *The Comforters* to provide a naturalistic explanation of Caroline's hallucinations; they are just that, the result of a psychological disorder, and in time they disappear. But she is suspicious of too much naturalism, like other Catholic novelists, and is not very interested in providing the plausible satisfactions of the traditional novel. The grandmother of Caroline's ex-lover Laurence, a delightful old lady living in the country, turns out to be running a gang of diamond-smugglers; some characters are involved in black magic; while Mrs Hogg, an ugly and vicious female retainer of Laurence's family, seems to be literally a witch, who becomes briefly invisible not long before she dies. *The Comforters* is a vigorous instance of the fiction of mixed literary modes, which will not be pinned down generically. It is, though, unified by its comic style and the cool, detached tone, which is close to that of Evelyn Waugh. In fact, Waugh wrote a very admiring review of *The Comforters* in which he generously allowed that Spark had handled the topic of aural hallucinations better than he had himself in his recently completed *The Ordeal of Gilbert Pinfold*. Waugh was somewhat puzzled by the metafictional dimension, and he described *The Comforters* as 'a very difficult book' but at the same time 'a thoroughly *enjoyable* work'.[3] His description still applies. *The Comforters* is a short but ambitious novel, and the various strands do to some extent become tangled. A 'mystery' is both a profound truth of religion and a crime story, and Spark is too inclined to force the parallels. A recent critic, Thomas Woodman, has remarked, 'The work of the novelist in *The Comforters* is tugged uneasily between an analogy with the trickeries of crime on the one hand and the providence of God on the other.'[4] Nevertheless, it was a dazzling fictional début.

In *Memento Mori* (1959) Muriel Spark moved into full maturity as a novelist, in a work which is tighter and more precise in plotting, and just as observant, cool, and witty. She presents a collection of elderly people, some of them upper middle class and comfortably off, still living in their own homes, and some of them poor, in a hospital for old people. Considered as comic realism, *Memento Mori* works wonderfully well, with a rich variety of characters—in some cases, perhaps, caricatures—and a faultless ear for speech. But the narrative voice laconically undoes any sense that the world of the novel is safe or comfortable, as in the opening words of chapter 5: 'Mrs Anthony knew instinctively that Mrs Pettigrew was a kindly woman. Her instinct was wrong.'

Spark's characters have all got to face death sooner rather than later; at the same time, they have very long memories, and the past weighs heavily upon them: the reverberations of a love affair in 1907, more than fifty years before, continue to impinge in the present. They are sharp-tongued and not very fond of each other. An element of fantasy, or of transcendent reality, intervenes when Dame Lettie Colston, one of the fittest and busiest of the old people, receives an anonymous telephone call telling her to remember she must die—the 'memento mori' of the title. Much later in the story she does die, murdered by a burglar. Meanwhile the calls have continued but cannot be traced. At first they are put down to a malicious hoaxer, until other people start getting the identical message but delivered in a great variety of voices and accents. The caller has to be Death himself, one of the most thoughtful of the characters reflects, and so it proves; the modern comedy of manners merges with the medieval morality play. Significantly it is not God but his agent Death who is the active force in the story as hoaxer and joker. Spark's religious assumptions pervade her painful comedy, where human beings are only too liable to act in foolish and selfish ways, even when death is approaching. The story is underpinned by a severe Augustinian, anti-humanistic form of Catholicism, emphasizing the gap between the human and the divine, and with little trace of a loving God. In this respect, though in no other, *Memento Mori* resembles an earlier Catholic novel, Greene's *Brighton Rock*. Otherwise, it is a work of fine comic invention, which is doubly disturbing; partly because the pervasive message, 'memento mori', reminds us that we too have to die, and partly because Spark's unrelentingly cool tone is a sign of the iciness at the heart of the book.

In his review of *The Comforters* Evelyn Waugh remarked, 'The only book I can think of which has any affinity with this is *Cards of Identity*. That was in many ways more elegantly written, but it had a basic futility which *The Comforters* escapes.' Waugh was referring to Nigel Dennis's *Cards of Identity* (1955), which made a great impact when it was published and ran into several impressions, though it now seems to have been forgotten. It is a comic fantasy with a strong thematic interest, as indicated by its title, which refers to the identity cards which everyone had to carry during the war and for some years afterwards, and more generally to the problem of identity in all its aspects, philosophical, psychological, and sociological. Dennis's story has elements both of Kafka and of *Alice in Wonderland*, and is rich in the peculiarly English form of surrealist humour that appeared in the Goon Shows on the radio, and later in Monty Python's Flying Circus on television. As Waugh said, Dennis writes very elegantly, and his underlying idea is a perfectly serious one: that human identity in the modern world has become so unstable that it can be changed without great difficulty. But some of the best comic effects are marginal to the theme.

The novel is set in an English country house, Hyde's Mortimer—'this sort of house was once a heart and centre of the national identity'—which was closed up during the war but is now restored to use. A body called the Identity Club has taken it over, and in order to find staff the Club's representatives abduct some of the local inhabitants and by unspecified but effective means transform them into the typical servants of the English country house, as we have met them in many novels, like the butler, the cook, and the eccentric gardener. The Club holds a meeting at Hyde's Mortimer, at which case-histories of identity change are read and discussed. Much of the novel consists of the texts of these papers. The first of them, 'The Case of the Co-Warden of the Badgeries', is a splendid satire on the monarchical trappings that survive in

English culture, but 'Secret Agent: Multiple Confessions and Singular Identities', about ex-communists writing their confessions in a monastery, is now very dated. On the other hand, 'Dog's Way: A Case of Multiple Sexual Misidentity', is rather ahead of its time. It deals light-heartedly with questions of androgyny and sexual ambiguity and confusion about gender, of a kind which did not become public issues until appreciably later. The meeting of the Club concludes with the performance of a Shakespeare play called *The Prince of Antioch or An Old Way to New Identity*, which is set down in full, in creditable pastiche of Shakespearean verse. *Cards of Identity* is an exuberant and ingenious fantasy. Though it is uneven, and some parts of it have not survived very well, it remains entertaining and intelligent enough to be worth making available again to a new generation of readers.

The first novels of Golding, Murdoch, and Spark, and Dennis's *Cards of Identity* (his second novel; his first, *Boys and Girls Come Out to Play*, had attracted little attention when it appeared in 1949) show that there was much more to the new fiction of the Fifties than the provincial realism, comic or sober, which is still sometimes thought of as dominating the period.

II

The insularity of the Movement was more of a provocative stance than a coherent attitude, but it was expressed in Amis's third novel, *I Like it Here* (1958), and in his notorious observation that no one wanted any more poems about foreign cities. Poems were written as specific rejoinders to it, such as Donald Davie's 'Via Portello', which is about Padua, and Charles Tomlinson's 'More Foreign Cities', which describes a number of beautiful foreign cities, most of them imaginary. Davie, though regarded from the beginning as a Movement poet, and for a time ready to see himself as one, was always opposed to the insularity and anti-modernism of Larkin and Amis. When Tomlinson's collection of poems, *Seeing is Believing*, appeared in 1958 Davie greeted it warmly, finding in Tomlinson an English poet whose masters were the great

French and American modernists, and who approached the art of poetry with proper seriousness. Tomlinson was born in 1927 and had published two pamphlets of poetry earlier in the Fifties, but his *Seeing is Believing* found no takers and it was brought out by an American publisher; the first British edition appeared in 1960. It is evident that Davie found in Tomlinson's work an encouragement to his own efforts to move out of the impasse of contemporary English poetics; he described Tomlinson as 'this most profound and original of all our postwar poets', adding that 'he refuses to join the silent conspiracy which now unites all the English poets from Robert Graves down to Philip Larkin, and all the critics, editors and publishers too, the conspiracy to pretend that Eliot and Pound never happened.'[5]

Davie may have been exaggerating in the excitement of his discovery; if pressed, I would still prefer Larkin to Tomlinson. Nevertheless, Tomlinson is a fine poet, who dwells with suave concentration on the beauty of the physical world and the power of art. As Davie implied, Tomlinson trained himself with Mallarmé and Laforgue, Pound and Eliot, Stevens and Williams. But further back there is Ruskin and the tradition that stemmed from him, of finding both truth and beauty in the patient and exact observation of the details of nature or architecture. Tomlinson, in an original fashion, traced links for himself between aestheticism and modernism. The resulting poetry is impressive and austerely elegant, though also, for my taste, rather coolly remote, suggestive of a latter-day Robert Bridges. Tomlinson has been accused of lacking human interest, and although I feel there is something in the charge, I must quote his rejoinder to it, in 'Paring the Apple':

> There are portraits and still-lifes
> And the first, because 'human'
> Does not excel the second, and
> Neither is less weighted
> With a human gesture, than paring the apple
> With a human stillness.

Tomlinson was one of several important English writers who emerged in the late 1950s and whose work, from the beginning, was on a different track from the writing generally associated with the decade. Two others, better known than Tomlinson, are the poet Ted Hughes and the dramatist Harold Pinter. Tomlinson and Hughes are equally removed from the Movement poetry of sharp, rational, formal comment on social or moral dilemmas; at the same time, they are antithetical to each other. Both poets, it is true, are more interested in nature than society, and might be called 'nature poets', but at the cost of stretching that elastic term beyond any possible usefulness. Hughes's début in *The Hawk in the Rain* (1957) revealed a poet with a powerful voice and a driving energy unparalleled in his generation. Hughes has a Cambridge degree in archaeology and anthropology, and is now Poet Laureate of the United Kingdom; nevertheless he has always kept himself apart from the main currents of English literary culture, and he reminds me of Eliot's remark that Wyndham Lewis combined the mind of a civilized man with the energy of the savage. In his poetry the energy is reined in by a strong formal control, with a palpable sense of strain.

In *The Hawk in the Rain* Hughes was still an immature and derivative poet, heavily indebted to Hopkins and Dylan Thomas, and it is a book of exciting promise rather than sustained actual achievement. It does, though, contain some successful poems, such as 'October Dawn', which moves a long way in a few lines, from the first slight frost of an October morning to the advent of a new Ice Age, when 'Mammoth and Sabre-tooth celebrate | Reunion'. Animals and other creatures are among Hughes's favourite subjects in this book and its more assured successor, *Lupercal* (1960), and frequently in his later work. He writes about hawks, a jaguar, a macaw, horses, cats, thrushes, a bull, an otter, a pig, a pike; but without any sentimental or friendly feelings towards them. What Hughes is most interested in is the alien aspect of the creatures and their actual or potential capacity for violence:

> Pigs have hot blood, they feel like ovens.
> Their bite is worse than a horse's—
> They chop a half-moon clean out.

Even thrushes are defamiliarized: 'Terrifying are the attent sleek thrushes on the lawn, | More coiled steel than living . . .'; later in the poem, Hughes refers to their 'bullet and automatic | Purpose'. Edward Larrissy has remarked on the way that here, and often elsewhere in his poetry, Hughes uses figurative language drawn from mechanical and industrial process; like, for instance, the 'drilling rain'.[6] Larrissy points out that such language undermines Hughes's seemingly objective stance towards natural objects; it recuperates nature in human terms, albeit in a radically unfamiliar fashion. In fact, Hughes's poetry contains a new-fashioned expression of what Ruskin condemned as the 'pathetic fallacy', in which natural objects or events are supposed to mirror or enact human dramas; in this case, reflecting the universality of violence, whether 'natural' or mechanical. It is illuminating to compare the way in which Tomlinson and Hughes have written about the same creature, a hawk. In Tomlinson's poem, 'How Still the Hawk' the human observer celebrates the beauty of the distant hawk as it first hovers and then swoops on its prey:

> Plummet of peace
> To him who does not share
> The nearness and the need . . .

Hughes's more dramatic 'Hawk Roosting' takes us into the consciousness of the hawk, emphasizing its absolute control over what it surveys and its total ruthlessness:

> I sit in the top of the wood, my eyes closed.
> Inaction, no falsifying dream
> Between my hooked head and hooked feet:
> Or in sleep rehearse perfect kills and eat . . .
>
> There is no sophistry in my body:
> My manners are tearing off heads—

The hawk provides an obvious symbol of the human, as gangster or dictator. If Tomlinson's view of nature is Ruskinian, Hughes's is Darwinian.

Hughes's fascination with violence and the physical or figurative links between its animal and human versions appears

in his interest in war; not the Second World War during which
he grew up, but the First World War, in which his father
fought. It is recalled in a number of raw but powerful poems
from *The Hawk in the Rain*: 'Bayonet Charge', 'Griefs for
Dead Soldiers', and 'Six Young Men'. Hughes impressively
brings together war and the violence of nature in the final lines
of 'Mayday on Holderness':

> The North Sea lies soundless. Beneath it
> Smoulder the wars: to heart-beats, bomb, bayonet.
> 'Mother, Mother!' cries the pierced helmet.
> Cordite oozings of Gallipoli,
>
> Curded to beastings, broached my palate,
> The expressionless gaze of the leopard,
> The coils of the sleeping anaconda,
> The nightlong frenzy of shrews.

Ted Hughes and Harold Pinter were both born in 1930, but
otherwise there is no obvious connection between the poet
from rural Yorkshire and the dramatist from the East End of
London. Yet there is a quality of sometimes shocking violence
in the writing of both men. To what extent one can think of a
playwright like Pinter as contributing to literature is itself
a problem. Traditionally drama has always been treated as
one of the three major fictive genres of literature, and from
Shakespeare to Shaw English plays could be read as dramatic
literature as well as enjoyed in the theatre (and indeed some
plays, like much nineteenth century poetic drama, never
reached the stage and perhaps were not intended for it). But
from the 1950s onwards many plays provide, on the page, no
more than a minimal notation for acting, where pauses and
silence are as important as the words. Drama in some of its
aspects is, indeed, a non-verbal art which can function by
movement and mime. Samuel Beckett, an Irishman who wrote
in French from the 1940s onwards, played a large part in this
shift. He was awarded the Nobel Prize for Literature, and
there can be no doubt that what he wrote, in drama and
fiction, was literature of a distinguished order. Beckett was a
master of language, but he was also a master of silences,

driven, in his late writing, by an impulse towards doing without words altogether.

Beckett's *Waiting for Godot* was a major theatrical success in London in 1955. Beckett was a late modernist, a friend and disciple of Joyce, and his play pressed home the modernist lesson that writing in prose can be as essentially poetic as in verse; in particular it showed that innovative and non-naturalistic work in the theatre could develop along different lines from the poetic drama of Eliot and Fry. Beckett's play posed absurdist and existentialist questions about the goals and nature of life, but it was also comic, drawing on the traditions of music-hall. It was also, in the relations between Lucky and Pozzo, disturbingly cruel. Beckett had an immediate and strong influence on young English dramatists, including Pinter.

Pinter understands words and knows how to use them; one of his greatest strengths is his ear for everyday speech and the repetitions and divagations and failures to respond that characterize so much ordinary conversation. At the same time, reading his plays I have a strong sense of how much they need theatrical performance to make dramatic sense, and to enforce the effects that are only hinted at on the page. Pinter deals in inexplicable menace, and his earliest plays show the actual or threatened disruption of a domestic situation. His one-act play *The Room* ends with a blind Negro being knocked down and savagely kicked; another one-acter, *The Dumb Waiter*, presents two hired killers waiting for their next assignment, passing the time in random conversation, and some comic business involving the restaurant upstairs. One of them, Gus, recalls the nasty mess a woman made after she was shot dead and wonders who cleared up afterwards. At the end of the play the other gunman, Ben, gets his orders via the speaking-tube from the restaurant; as the curtain falls he is about to shoot Gus.

Pinter's first full-length play, *The Birthday Party*, is set in a boarding-house in a seaside town. It focuses on the birthday party being arranged for one of the boarders, Stanley. The party proves to be a threatening occasion, and the play ends with Stanley being taken away by two other boarders, Goldberg and McCann, to a sinister fate. *The Birthday Party* was a sustained exploration of what was to become Pinter's characteristic material: the juxtaposition of domesticity and menace. *The Caretaker*, performed in 1960, was a major success. Like the others, it takes place in an enclosed space: in this case, a room cluttered with junk. It is supposed to be somewhere in London, just as *The Dumb Waiter* was supposedly set in Birmingham, but in reality both plays exist in a wasteland as desolate as any of Beckett's. There are only three characters: the elderly tramp Davies, and the two brothers Aston and Mick. In so far as Davies upsets the way of life of the two brothers, he is a further instance of the invading outsider who appeared in Pinter's earlier plays, though here the disruption is not violent. Davies behaves with contempt towards the brain-damaged Aston who had befriended him, and in the end he is turned out. The play makes a chilling though often comic exploration of the impossibility of communication between human beings, whether on the level of talk or of deeper understanding.

Davies, always talking about the 'papers' in Sidcup that would legitimize his existence if only he could get hold of them, can be seen as a type of the refugee or displaced person, who were noticeable features of the postwar world. It has to be said that Pinter shows no direct awareness of politics and history and would probably reject so simple an interpretation. Yet it seems inevitable that he was influenced by the war and its aftermath. John Russell Taylor has remarked of the threatening knock on the door or other invasion of domestic space: 'Pinter, himself a Jew, grew up during the war, precisely the time when the menace inherent in such a situation would have been, through the medium of the cinema or of radio, most imaginatively present to any child, and particularly perhaps a Jewish child.'[7] Pinter has always refused to elucidate his intentions, and most discussion of his plays relies heavily on the ideas of generalized menace and non-communication. The plays have been explicated in a predictable fashion rather than given serious critical discussion; though this may be inevitable in dealing with work that, like much recent drama, is only 'literary' in a partial and problematical sense. My own conclusion is that Pinter has a strong but narrow talent, and is inclined to mystification for its own sake; he can focus on only a limited number of dramatic possibilities.

The emerging English writers of the 1940s and 1950s had lived through a period of unprecedented displacement, atrocity, and destruction. Orwell and Koestler, who had seen something of these things at first hand, tried to respond to them, as Sartre and Camus did in France. And so, perhaps, did Beckett, who had served in the French Resistance. But for some time the English were capable only of an anaesthetized or evasive response: 'How dare we now be anything but numb?' as Donald Davie wrote. The new English writing, despite the cult of anger, was comparatively peaceful, notwithstanding the fights in *Saturday Night and Sunday Morning* and Jimmy Porter's savage verbal tirades in *Look Back in Anger*. It would be a mistake to look for any direct reflection of recent history in the violence and cruelty that appear, in different ways, in the work of Hughes and Pinter towards the end of the 1950s, or in the romantic agonies and mutilations of Durrell's *Alexandria Quartet*. But they indicate a shift of sensibility.

9

Into the Sixties

In 1953 John Wain said that the task for new writers was to consolidate the great literary achievements of the earlier years of the century and hold back from further advances for the time being. Some of them interpreted 'consolidation' as rejection, but by 1960 what was called the reaction against experiment seemed to be coming to an end. In a broader sense, too, the Fifties were a period of consolidation, as when a convalescent has to consolidate his resources until his strength returns. The slow recovery from the trauma of war was accompanied by insular and formally conservative kinds of writing, though within those modes some excellent work was done. In the national culture a perceptible increase in material prosperity and general well-being had arrived by the late Fifties, and was reflected in the Conservative victory in the General Election of 1959. A couple of years before, the wily patrician Prime Minister, Harold Macmillan, had told the British people, 'You've never had it so good', cynically and significantly using an American electoral slogan. The patient was on his feet again, perhaps ready for a little adventure.

Realism is a major mode of literary consciousness and production, and is likely to remain so. Certainly, realistic fiction was still very popular in 1960, when Amis published *Take a Girl Like You*, a novel about virginity which is consciously indebted to Richardson's *Clarissa*. It documented the cultural transition from a time when nice girls don't to a time when they might, and quite soon, assisted by the contraceptive pill, certainly will. In the same year, Lynne Reid Banks's *The L-Shaped Room* showed what happened to one nice girl who did. Two other first novels published in 1960, David Lodge's *The Picturegoers* and David Storey's *This Sporting Life*, were

works of four-square realism, one set in South London sub-
urbia, the other in a northern town. But the later development of both writers took them further, into metafiction and
mythopoiea, and others moved in the same direction. Iris
Murdoch's *A Severed Head* (1961) was at odds in most respects with the norms of Fifties realism; it presented well-off
people, living in the acceptable parts of West London, and
although it exemplifies Murdoch's obsessive concern with the
topography of the capital, the descriptive writing readily runs
to purple passages. The story develops into a rich mixture of
myth and violence and sexual complication and deviance;
a sinister anthropologist called Honor Klein dominates the
action like a primitive deity.

In 1962 'experiment' came noticeably back on the literary
scene, in Doris Lessing's remarkable long novel, *The Golden
Notebook*, an elaborate metafictional exploration of the pro-
cesses of imaginative writing, and B. S. Johnson's Sterneian
first novel, *Travelling People*. And in that year A. Alvarez
edited a Penguin anthology, *The New Poetry*, claiming in his
introduction that English poets needed to write in full consciousness of the violence and horror of the modern world,
facing both its collective manifestations, revealed in recent
history, and those uncovered in the individual by psychoanalysis: 'it is hard to live in an age of psychoanalysis and feel
oneself wholly detached from the dominant public savagery.'
Alvarez mildly mocked the *New Lines* poets for their gentility
and timidity, unfavourably comparing Larkin's 'At Grass' with
Hughes's Lawrentian 'A Dream of Horses'; he represented
the Movement poets in his anthology, but prefaced it with a
selection from the American confessional poets whom he saw
as admirably ready to embark on the perilous encounter with
inner and outer violence: John Berryman, Robert Lowell,
Anne Sexton, and Sylvia Plath (three of whom were to take
their own lives). 1962 was the year of the Cuban Missile crisis,
when for a few days the world trembled on the brink of the
nuclear war that had inspired much futurological writing.

I conclude with an event in 1960 which had wide cultural
and literary implications. English writers had long been

hampered by the laws relating to obscenity, which meant that common human actions could not be described, and common words could not be printed. (The ban on writing 'fuck' and its extensions involved writers in much tedious ingenuity to suggest it, particularly in reproducing working-class or soldiers' speech.) During the 1950s the law was strictly applied, and a number of publishers were fined for breaking it; the crowning absurdity came from the magistrates at Swindon who ordered copies of Boccaccio's *Decameron* to be destroyed. As it stood, the law made no distinction between avowedly pornographic works and writing that was intended to be literary. Eventually, after much pressure from informed public opinion, the law was changed to admit a plea of literary merit to be made in defence.[1]

After the new and somewhat more liberal law was introduced, Penguin Books resolved to test it forthwith. In 1960 they brought out a new impression of D. H. Lawrence's novels to mark the thirtieth anniversary of his death. Among them they included the long-banned *Lady Chatterley's Lover*, a gentle moral fable which Lawrence wrote shortly before his death, and which is not among his greatest books. Its earnest frankness of sexual description and language meant that it could not be published unexpurgated in Britain, and the well-thumbed copies of the Continental edition that circulated clandestinely gave the book a mythic notoriety. The trial of Penguin Books in October 1960 for publishing an allegedly obscene book was a ritual event of the greatest fascination. The new Obscene Publications Act meant that expert witnesses could be called by the defence to testify to the literary merit of a work, and an illustrious succession of them appeared, including writers, critics, academics, journalists, and clergymen. The youngest witness was the convent-educated daughter of a Catholic literary family, just 21 and a recent Cambridge graduate in English. She testified that she had read the unexpurgated version of Lawrence's novel and found it an improving book.

The prosecuting counsel for the Crown, Mervyn Griffith-Jones, QC, struck an unfortunate note in his opening address, when he asked the jury, 'Is it a book that you would have lying around in your own house? Is it a book that you would even wish your wife or your servants to read?' Three of the jurors were in fact women, and Mr Griffith-Jones seemed not to be living in the same world as the jury or the witnesses or the potential readers of the novel. He was certainly baffled by some of the claims made by the defence witnesses, as when the Bishop of Woolwich said of Lawrence's treatment of sex, 'I think Lawrence tried to portray this relation as in a real sense something sacred, as in a real sense an act of holy communion'; or when Richard Hoggart affirmed that *Lady Chatterley's Lover* was 'highly moral and not degrading of sex'. Some rather extravagant things were said by the defence witnesses about this fairly minor text, but the languages of law and literary criticism are very different discourses, and high claims had to be advanced if any point was to stick. The prosecution called no experts, though it is said that Evelyn Waugh volunteered to appear, and at the end of the trial the publisher was acquitted. Lawrence's notorious novel could be legally printed.

Penguin Books had taken a rational gamble in publishing *Lady Chatterley's Lover* and inviting a test case in law; if they had lost, they would have had to pay a fine plus costs, and lose any copies of the book that had been printed. By winning, they did extremely well; six weeks after the trial ended two million copies had been printed and sold.[2] Penguin Books were associated with the cultural ideals of wartime England, which they had done so much to establish, and their success—aided by the decent, liberal-minded intellectuals who had lined up in their defence—was an unexpected victory for those whom Michael Frayn called the Herbivores, who had been on the defensive since the return of Conservative rule.

After the reticences and the sober realism, the moderate anger and mild ironies of the Fifties, the Sixties let it all hang out, with sex and drugs and rock-and-roll. Philip Larkin, in his ruefully retrospective poem, 'Annus Mirabilis', claimed, in what has since become a proverbial utterance, that sexual intercourse began in 1963, 'Between the end of the *Chatter-*

ley ban | And the Beatles' first LP'. In America, Robert Zimmerman, who had become Bob Dylan as a tribute to the Welsh poet, sang 'The times they are a-changing', 'Lay, lady, lay', and 'Everybody must get stoned'.

Notes

Chapter 1: *Blackout to Blitz*

1. Roy Fuller, *The Strange and the Good: Collected Memoirs* (London, 1989), 199.
2. Derek Stanford, *Inside the Forties: Literary Memoirs 1937–1957* (London, 1977), 5.
3. Richard Hillary, *The Last Enemy* (London, 1961), 35.
4. Desmond Graham, *Keith Douglas* (London, 1974), 79.
5. Clive Ponting, *1940: Myth and Reality* (London, 1990), 92.
6. Paul Fussell, *Wartime: Understanding and Behaviour in the Second World War* (Oxford, 1989), 275.
7. S. Schimanski and H. Treece (eds.), *Leaves in the Storm: A Book of Diaries* (London, 1947), 56.
8. See 'Auden/Greene' in B. Bergonzi, *Reading the Thirties* (London, 1978).
9. George Orwell, *Collected Essays, Journalism and Letters*, iii (Harmondsworth, 1970), 271.
10. ibid. 275.
11. *Scrutiny*, Dec. 1942.
12. Hillary, *Last Enemy*, 1.
13. Julian Symons, *Notes from Another Country* (London, 1972), 84.
14. Fuller, *Strange and the Good*, 200.
15. Orwell, *Essays*, ii. 273.
16. *Horizon*, Dec. 1940.
17. Louis MacNeice, *Selected Prose* (Oxford, 1990), 103.
18. ibid. 118.
19. Graham Greene, 'At Home', *The Lost Childhood and Other Essays* (Harmondsworth, 1962), 223.
20. Graham Greene, *Ways of Escape* (Harmondsworth, 1982), 84–8.
21. *Horizon*, May 1941.
22. Orwell, *Essays*, ii. 74.

Chapter 2: *Writers on an Island*

1. Arthur Koestler, *The Yogi and the Commissar* (London, 1945), 25.

2. ibid. 36–7.
3. Robert Hewison in A. T. Tolley (ed.), *John Lehmann: A Tribute* (Ottawa, 1987), 116.
4. Fussell, *Wartime*, 88.
5. Julian Maclaren-Ross, *Memoirs of the Forties* (Harmondsworth, 1984), 25.
6. Walter Allen, *Tradition and Dream* (Harmondsworth, 1965), 267.
7. Evelyn Waugh, *Diaries 1911–1965* (Harmondsworth, 1979), 608.
8. Fussell, *Wartime*, 228.
9. Maclaren-Ross, *Memoirs of the Forties*, 232.
10. Elizabeth Bowen, quoted in R. Blyth (ed.), *Components of the Scene* (Harmondsworth, 1966), 389.
11. Orwell, *Essays*, ii. 108.
12. MacNeice, *Selected Prose*, 127.
13. Orwell, *Essays*, ii. 75–6.
14. Koestler, *Yogi and the Commissar*, 67.

Chapter 3: *Poets at Home and Abroad*

1. For a sympathetic account of the movement, see Andrew Crozier, 'Styles of the Self: the New Apocalypse and 1940s Poetry' in D. Mellor (ed.), *A Paradise Lost* (London, 1987), 114–16.
2. Donald Davie, *The Poet in the Imaginary Museum* (Manchester, 1977), 68.
3. G. S. Fraser, *Essays on Twentieth Century Poets* (Leicester, 1977), 201.
4. Henry Treece, *How I See Apocalypse* (London, 1946), 176.
5. Philip Larkin, *The North Ship* (2nd edn., London, 1966), 9.
6. William Empson, *Argufying* (London, 1987), 382–6.
7. Julian Symons, *Critical Observations* (London, 1981), 37.
8. *Horizon*, Jan. 1941.
9. D. Graham (ed.), *Keith Douglas: A Prose Miscellany* (Manchester, 1985), 119–20.
10. ibid. 17.
11. Vernon Scannell, *Not Without Glory* (London, 1976), 43.
12. I have said more about the Cairo poets in 'Poets of the 1940s', Bergonzi, *The Myth of Modernism and Twentieth Century Literature* (Brighton, 1986); and 'Poetry of the Desert War' in V. Bell and L. Lerner (eds.), *On Modern Poetry: Essays Presented to Donald Davie* (Nashville, Tenn., 1988). Collected editions of these poets include Douglas, *Complete Poems*

(Oxford, 1978); Durrell, *Collected Poems 1931–1974* (London, 1980); Fraser, *Poems* (Leicester, 1981); Spencer, *Collected Poems* (Oxford, 1981).

13. Fuller, *Strange and the Good*, 239.
14. Alan Ross, *Blindford Games* (London, 1986), 240.

Chapter 4: *The Wake of War*

1. A. T. Tolley, *Poetry of the Forties* (Manchester, 1985), 197.
2. Michael Frayn, 'Festival', M. Sissons and P. French (eds.), *Age of Austerity 1945–51* (Harmondsworth, 1964), 331–2.

Chapter 5: *Sequences*

1. Interview, *Twentieth Century*, July 1961, 53.
2. Allen, *Tradition and Dream*, 76–7.
3. Durrell, H. T. Moore (ed.), *The World of Lawrence Durrell* (New York, 1964), 231.
4. ibid. 138–9.
5. Rubin Rabinowitz, *The Reaction against Experiment in the English Novel 1950–1960* (New York, 1967), 155.
6. Stephen Wall, *London Magazine*, Apr. 1964.
7. See Bergonzi, *Situation of the Novel* (2nd edn., London, 1979), 118–33, 239–41.

Chapter 6: *Anger and the Empirical Temper*

1. Kate Whitehead, *The Third Programme: A Literary History* (Oxford, 1989), 200.
2. Allen, *Tradition and Dream*, 300–1.
3. *Spectator*, 15 Oct. 1954.
4. Malcolm Bradbury, *No, Not Bloomsbury* (London, 1987), 183.
5. T. Maschler (ed.), *Declaration* (London, 1957), 23.
6. Kingsley Amis, *Memoirs* (London, 1991), 155.
7. Raymond Williams, *The Long Revolution* (London, 1961), 285.
8. John Russell Taylor, *Anger and After: A Guide to the New British Drama* (Harmondsworth, 1963), 38.
9. Colin MacInnes, *England, Half English* (Harmondsworth, 1966), 204.
10. Davie, *Poet in the Imaginary Museum*, 144.
11. For more on Davie, see 'Pound and Donald Davie' and 'Davie, Larkin and the State of England' in Bergonzi, *Myth of Modernism*.
12. *Independent Magazine*, 16 Mar. 1991.

13. John Wain, *Preliminary Essays* (London, 1957), 182–3.
14. *Declaration*, 32.

Chapter 7: *The Myth Kitty*

1. Allen, *Tradition and Dream*, 286.
2. D. Wright (ed.), *Longer Contemporary Poems* (Harmondsworth, 1966), 10.

Chapter 8: *Contrary Voices*

1. M. Bradbury, *No, Not Bloomsbury*, 255.
2. R. Whitaker, M. Bradbury and D. Palmer (eds.), *The Contemporary English Novel* (London, 1979), 163.
3. D. Gallagher (ed.), *The Essays, Articles and Reviews of Evelyn Waugh* (London, 1983), 518–19.
4. Thomas Woodman, *Faithful Fictions: The Catholic Novel in British Literature* (Milton Keynes, 1991), 126.
5. Davie, *Poet in the Imaginary Museum*, 66–7.
6. Edward Larrissy, *Reading Twentieth Century Poetry* (Oxford, 1990), 126.
7. Taylor, *Anger and After*, 288.

Chapter 9: *Into the Sixties*

1. See 'Censorship and the Novel' in George Greenfield, *Scribblers for Bread: Aspects of the English Novel since 1945* (London, 1989).
2. ibid. 100.

Chronology

1939   Germany occupies Czechoslovakia. End of Spanish Civil War. Russo-German pact. German invasion of Poland and outbreak of Second World War. Deaths of Yeats, Freud, and Ford Madox Ford.

Roy Fuller, *Poems*. Christopher Isherwood, *Goodbye to Berlin*. James Joyce, *Finnegans Wake*. Louis MacNeice, *Autumn Journal*. George Orwell, *Coming Up for Air*. Dylan Thomas, *The Map of Love*. *The New Apocalypse* (anthology).

1940   Churchill prime minister. Fall of France. Evacuation from Dunkirk. Italy enters war on side of Germany. Battle of Britain and bombing of London and other cities.

W. H. Auden, *Another Time*. T. S. Eliot, *East Coker*. Graham Greene, *The Power and the Glory*. Arthur Koestler, *Darkness at Noon*. W. B. Yeats, *Last Poems and Plays*.

1941   German invasion of Russia. Japanese attack on Pearl Harbor. USA at war with Germany, Italy, and Japan. Deaths of Joyce and Virginia Woolf.

Auden, *New Year Letter*. Joyce Cary, *Herself Surprised*. Patrick Hamilton, *Hangover Square*. Orwell, *The Lion and the Unicorn*. Rex Warner, *The Aerodrome*. Virginia Woolf, *Between the Acts*. *The White Horseman* (anthology).

1942   Hitler announces destruction of European Jews. Fall of Singapore to the Japanese. Battle of El Alamein.

Cary, *To be a Pilgrim*. Fuller, *The Middle of a War*. Richard Hillary, *The Last Enemy*. Sidney Keyes, *The Iron Laurel*. Alun Lewis, *Raiders' Dawn*. Evelyn Waugh, *Put Out More Flags*.

1943   German defeat at Stalingrad. Conquest of Tunisia and Sicily. Invasion and surrender of Italy.

Lawrence Durrell, *A Private Country*. David Gascoyne, *Poems 1937–42*. Henry Green, *Caught*. Greene, *The Ministry of Fear*. James Hanley, *No Directions*. Hamilton, *Slaves of Solitude*. Alun Lewis, *The Last Inspection*.

1944   Allies invade France. Flying bomb and rocket attacks on London.

Cary, *The Horse's Mouth*. Cyril Connolly (Palinurus), *The Unquiet Grave*. Eliot, *Four Quartets*. L. P. Hartley, *The Shrimp and the Anemone*. Julian Maclaren-Ross, *The Stuff to Give the Troops*.

1945   Death of Roosevelt; Truman president of USA. End of war in Europe. Labour government elected; Attlee prime minister. Atomic bombs on Japan and Japanese surrender.

Green, *Loving*. Koestler, *The Yogi and the Commissar*. Philip Larkin, *The North Ship*. Orwell, *Animal Farm*. Edith Sitwell, *The Song of the Cold*. Waugh, *Brideshead Revisited*.

1946   First assembly of United Nations. Death of H. G. Wells.

Keith Douglas, *Alamein to Zem Zem*.

Robert Graves, *Poems 1938–45*. Larkin, *Jill*. Edwin Muir, *The Voyage*. Dylan Thomas, *Deaths and Entrances*. R. S. Thomas, *The Stones of the Field*.

1947   Nationalization of mines. Marshall Plan for American aid to Europe. Independence of India.

Larkin, *A Girl in Winter*. Malcolm Lowry, *Under the Volcano*.

1948   Assassination of Gandhi. Communist takeover in Czechoslovakia. Russian blockade of Berlin. State of Israel founded.

Greene, *The Heart of the Matter*. Graves, *The White Goddess*. F. R. Leavis, *The Great Tradition*.

1949   Communist government in China. North Atlantic Treaty Organization set up. Pound devalued.

Elizabeth Bowen, *The Heat of the Day*. Eliot, *The Cocktail Party* (first performance). Aldous Huxley, *Ape and Essence*. MacNeice, *Collected Poems 1925–48*. Muir, *The Labyrinth*. William Sansom, *The Body*. Angus Wilson, *The Wrong Set*.

1950   Korean war. Deaths of Shaw and Orwell.

Auden, *Collected Shorter Poems*. George Barker, *The Dead Seagull*; *True Confession*. William Cooper, *Scenes from Provincial Life*. Doris Lessing, *The Grass is Singing*. Wilson, *Such Darling Dodos*.

1951 Conservative government elected; Churchill prime minister. Death of Wittgenstein.

Keith Douglas, *Collected Poems*. Greene, *The End of the Affair*. Anthony Powell, *A Question of Upbringing*. C. P. Snow, *The Masters*.

1952 Death of George VI; accession of Elizabeth II. Eisenhower elected president of USA.

David Jones, *The Anathemata*. Leavis, *The Common Pursuit*. Powell, *A Buyers' Market*. Dylan Thomas, *Collected Poems*. Waugh, *Men at Arms*. Wilson, *Hemlock and After*.

1953 Death of Stalin. End of Korean War. Coronation of Queen Elizabeth II. Everest climbed. Death of Dylan Thomas.

Hartley, *The Go-Between*. Elizabeth Jennings, *Poems*. John Wain, *Hurry On Down*. Wittgenstein, *Philosophical Investigations*.

1954 Geneva Conference on Indo-China. Outbreak of war in Algeria.

Kingsley Amis, *Lucky Jim*. William Golding, *Lord of the Flies*. Thom Gunn, *Fighting Terms*. Wyndham Lewis, *Self Condemned*. Iris Murdoch, *Under the Net*.

1955 Eden prime minister.

Donald Davie, *Brides of Reason*. Nigel Dennis, *Cards of Identity*. William Empson, *Collected Poems*. Golding, *The Inheritors*. W. S. Graham, *The Nightfishing*. Greene, *The Quiet American*. Larkin, *The Less Deceived*. Hugh MacDiarmid, *In Memoriam James Joyce*. Powell, *The Acceptance World*. R. S. Thomas, *Song at the Year's Turning*. Waugh, *Officers and Gentlemen*.

1956 Revolution in Hungary and Russian occupation. Anglo-French attack on Egypt.

Muir, *One Foot in Eden*. John Osborne, *Look Back in Anger* (first performed). Samuel Selvon, *The Lonely Londoners*. Angus Wilson, *Anglo-Saxon Attitudes*. Colin Wilson, *The Outsider*. *New Lines* (anthology).

1957 Macmillan prime minister. Russian sputnik in orbit. Death of Wyndham Lewis.

John Braine, *Room at the Top*. Davie, *A Winter Talent*. Durrell, *Justine*. Gunn, *The Sense of Movement*. Richard Hoggart, *The Uses of Literacy*. Ted Hughes, *The Hawk in the Rain*. Frank Kermode, *Romantic Image*. Colin MacInnes, *City of Spades*. Osborne, *The Entertainer* (first performed). Muriel Spark, *The Comforters*. Waugh, *The Ordeal of Gilbert Pinfold*.

1958 Overthrow of Fourth Republic in France; de Gaulle back in power. First Aldermaston march.

John Betjeman, *Collected Poems*. Shelagh Delaney, *A Taste of Honey* (first performed). Durrell, *Balthazar*, *Mountolive*. Harold Pinter, *The Birthday Party* (first performed). Alan Sillitoe, *Saturday Night and Sunday Morning*. Raymond Williams, *Culture and Society*.

1959 Castro takes power in Cuba. Death of Muir.

Malcolm Bradbury, *Eating People is Wrong*. Golding, *Free Fall*. Colin MacInnes, *Absolute Beginners*. Spark, *Memento Mori*. Sillitoe, *The Loneliness of the Long Distance Runner*. Keith Waterhouse, *Billy Liar*.

1960 Kennedy elected president of USA. United Nations intervention in Congo. The 'Lady Chatterley' Trial.

Amis, *Take a Girl Like You*. Lynne Reid Banks, *The L-Shaped Room*. Stan Barstow, *A Kind of Loving*. Durrell, *Clea*. Hughes, *Lupercal*. David Lodge, *The Picturegoers*. MacInnes, *Mr Love and Justice*. Pinter, *The Caretaker* (first performed). David Storey, *This Sporting Life*. Charles Tomlinson, *Seeing is Believing*.

For Further Reading

(The place of publication is London unless otherwise stated.)

The War Years: (i) *Political and Social History*

Paul Addison, *The Road to 1945: British Politics and the Second World War* (1975); Correlli Barnett, *The Audit of War: The Illusion and Reality of Britain as a Great Nation* (1986); Angus Calder, *The People's War: Britain 1939–45* (1969); Constantine Fitzgibbon, *The Blitz* (1957); Norman Longmate, *How We Lived Then: A History of Everyday Life During the Second World War* (1971); Clive Ponting, *1940: Myth and Reality* (1990).

The War Years: (ii) *Literary and Cultural Studies*

Walter Allen, *Tradition and Dream: A Critical Survey of British and American Fiction from the 1920s to the Present Day* (1964) (this work is invaluable for the whole period under discussion); Paul Fussell, *Wartime: Understanding and Behaviour in the Second World War* (Oxford, 1989); Robert Hewison, *Under Siege: Literary Life in London 1939–45* (1977); Julian Maclaren-Ross, *Memoirs of the Forties* (1965); David Mellor (ed.), *A Paradise Lost: The Neo-Romantic Imagination in Britain 1935–55* (1987) (primarily concerned with painting but includes accounts of poetry and film); Andrew Sinclair, *War Like a Wasp: The Lost Decade of the Forties* (1989); Derek Stanford, *Inside the Forties: Literary Memoirs 1937–1957* (1977).

After 1945: (i) *Political and Social History*

Paul Addison, *Now the War is Over: A Social History of Britain 1945–51* (1985); Noel Annan, *Our Age: the Generation that Made Post-War Britain* (1990); Arthur Marwick, *British Society since 1945* (Harmondsworth, 1982; 2nd edn., 1990); John Montgomery, *The Fifties* (1965); Michael Sissons and Philip French (eds.), *Age of Austerity 1945–51* (1963).

After 1945: (ii) *Literary and Cultural Studies*

Bryan Appleyard, *The Pleasures of Peace: Art and Imagination in*

Post-War Britain (1989); Bernard Bergonzi, *The Situation of the Novel* (1970; 2nd edn., 1979); Malcolm Bradbury, *No, Not Bloomsbury* (1987); James Gindin, *Postwar British Fiction* (1962); George Greenfield, *Scribblers for Bread: Aspects of the English Novel since 1945* (1989); Robert Hewison, *In Anger: Culture in the Cold War* (1981); Arthur Marwick, *Culture in Britain Since 1945* (Oxford, 1991); Alan Sinfield, *Literature, Politics and Culture in Post-War Britain* (Oxford, 1989) and (ed.), *Society and Literature 1945–1970* (1983); George Watson, *British Literature Since 1945* (1991); Kate Whitehead, *The Third Programme: A Literary History* (Oxford, 1989).

1. *Blackout to Blitz*

Humphrey Carpenter, *W. H. Auden: A Biography* (1981); Michael Shelden, *Friends of Promise: Cyril Connolly and the World of Horizon* (1989); Peter Ackroyd, *T. S. Eliot* (1984); Helen Gardner, *The Composition of Four Quartets* (1978); Norman Sherry, *The Life of Graham Greene: Vol. 1, 1904–1939* (1989); Roger Sharrock, *Saints, Sinners and Comedians: The Novels of Graham Greene* (Tunbridge Wells, 1984); Iain Hamilton, *Koestler: A Biography* (1982); Bernard Crick, *George Orwell: A Life* (1980); Martin Stannard, *Evelyn Waugh: The Early Years 1903–1939* (1986), *Evelyn Waugh: No Abiding City 1939–1966* (1992), (ed.), *Evelyn Waugh: The Critical Heritage* (1984).

2. *Writers on an Island*

Victoria Glendenning, *Elizabeth Bowen: Portrait of a Writer* (1977); Malcolm Foster, *Joyce Cary: A Biography* (1969); Rod Mengham, *The Idiom of the Time: The Writings of Henry Green* (Cambridge, 1982); Nigel Jones, *Through a Glass Darkly: The Life of Patrick Hamilton* (1991); A. T. Tolley (ed.), *John Lehmann: A Tribute* (Ottawa, 1987); John Pikoulis, *Alun Lewis: A Life* (Bridgend, 1984); Quentin Bell, *Virginia Woolf: A Biography*, 2 vols. (1972).

3. *Poets at Home and Abroad*

Vernon Scannell, *Not Without Glory: Poets of the Second World War* (1976); Francis Scarfe, *Auden and After: The Liberation of Poetry 1930–1941* (1942); Linda M. Shires, *British Poetry of the Second World War* (1985); Derek Stanford, *The Freedom of Poetry: Studies in Contemporary Verse* (1947); A. T. Tolley, *The Poetry of the Forties* (Manchester, 1985); Henry Treece, *How I See Apocalypse* (1946).

Desmond Graham, *Keith Douglas 1920–1944: A Biography* (1974); Barbara Coulton, *Louis MacNeice in the BBC* (1980); Robyn Marsack, *The Cave of Making: The Poetry of Louis MacNeice* (Oxford, 1982); John Pearson, *Façade: Edith, Osbert and Sacheverell Sitwell* (1978); Paul Ferris, *Dylan Thomas* (1977).

4. The Wake of War

Jean Hartley, *Philip Larkin, the Marvell Press and Me* (Manchester, 1989); Andrew Motion, *Philip Larkin* (1982); Peter Faulkner, *Angus Wilson: Mimic and Moralist* (1980).

5. Sequences

G. S. Fraser, *Lawrence Durrell: A Study*, (2nd edn., 1973); Peter Bien, *L. P. Hartley* (1963); Lorna Sage, *Doris Lessing* (1983); Jeffrey Myers, *Wyndham Lewis* (1980); Hilary Spurling, *Handbook to Anthony Powell's Music of Time* (1977); Neil McEwan, *Anthony Powell* (1991); Suguna Ramanathan, *The Novels of C. P. Snow: A Critical Introduction* (1982).

6. Anger and the Empirical Temper

Tom Maschler (ed.), *Declaration* (1957); Blake Morrison, *The Movement: English Poetry and Fiction of the 1950s* (Oxford, 1980); Rubin Rabinowitz, *The Reaction against Experiment in the English Novel 1950–1960* (New York, 1967); Harry Ritchie, *Success Stories: Literature and the Media in England 1950–1959* (1990); John Russell Taylor, *Anger and After: A Guide to the New British Drama* (1962). Richard Bradford, *Kingsley Amis* (1989); John McDermott, *Kingsley Amis: An English Moralist* (1989); George Dekker (ed.), *Donald Davie and the Responsibilities of Literature* (Manchester, 1983); William Walsh, *D. J. Enright: Poet of Humanism* (Cambridge, 1974); Simon Trussler, *The Plays of John Osborne: An Assessment* (1969); Stanley S. Atherton, *Alan Sillitoe: A Critical Assessment* (1979).

7. The Myth Kitty

Mark Kinkead-Weekes and Ian Gregor, *William Golding: A Critical Study* (2nd edn., 1984); Martin Seymour-Smith, *Robert Graves: His Life and Work* (1982); David Blamires, *David Jones: Artist and Writer* (Manchester, 1971); René Hague, *A Commentary on The Anathemata of David Jones* (Wellingborough, 1977); Douglas Day, *Malcolm Lowry: A Biography* (1974); Alan Bold, *MacDiarmid* (1988); Peter Butter, *Edwin Muir: Man and Poet* (Edinburgh, 1966).

8. *Contrary Voices*

Deirdre Bair, *Samuel Beckett: A Biography* (1978); Keith Sagar, *The Art of Ted Hughes* (2nd edn., Cambridge, 1978); Terry Gifford and Neil Roberts, *Ted Hughes: A Critical Study* (1981); Elizabeth Dipple, *Iris Murdoch: Work for the Spirit* (1982); Deborah Johnson, *Iris Murdoch* (Brighton, 1987); Ronald Hayman, *Harold Pinter* (1980); Martin Esslin, *Harold Pinter: A Study of His Plays* (1977); Peter Kemp, *Muriel Spark* (1974).

9. *Into the Sixties*

C. H. Rolph (ed.), *The Trial of Lady Chatterley* (Harmondsworth, 1961).

Index

(Dates of birth and death are shown for writers of the period whose work is discussed or cited in the text.)